DEBATING CLIMATE ETHICS

DEBATING ETHICS

General Editor
Christopher Heath Wellman
Washington University of St. Louis

Debating Ethics is a series of volumes in which leading scholars defend opposing views on timely ethical questions and core theoretical issues in contemporary moral, political, and legal philosophy.

Debating the Ethics of Immigration
Is There a Right to Exclude?
Christopher Heath Wellman and Philip Cole

Debating Brain Drain
May Governments Restrict Emigration?
Gillian Brock and Michael Blake

Debating Procreation
Is It Wrong to Reproduce?
David Benatar and David Wasserman

Debating Climate Ethics
Stephen Gardiner and David Weisbach

Debating
Climate Ethics

STEPHEN M. GARDINER
AND DAVID A. WEISBACH

OXFORD
UNIVERSITY PRESS

OXFORD

UNIVERSITY PRESS

Oxford University Press is a department of the University of Oxford. It furthers the University's objective of excellence in research, scholarship, and education by publishing worldwide. Oxford is a registered trade mark of Oxford University Press in the UK and certain other countries.

Published in the United States of America by Oxford University Press
198 Madison Avenue, New York, NY 10016, United States of America.

Library of Congress Cataloging-in-Publication Data
Names: Gardiner, Stephen Mark, author.
Title: Debating climate ethics / Stephen M. Gardiner, David A. Weisbach.
Description: New York: Oxford University Press, 2016. | Series: Debating ethics
| Includes bibliographical references and index.
Identifiers: LCCN 2015046545| ISBN 978-0-19-999648-3 (pbk.: alk. paper) |
ISBN 978-0-19-999647-6 (hardcover : alk. paper) | ISBN 978-0-19-061476-8
(ebook (epub)) | ISBN 978-0-19-999649-0 (ebook (updf))
Subjects: LCSH: Environmental ethics. | Climatic changes—Moral and
ethical aspects.
Classification: LCC GE42 .G36 2016 | DDC 179/.1—dc23 LC record available
at http://lccn.loc.gov/2015046545

3 5 7 9 8 6 4 2

Printed by Webcom, Canada

CONTENTS

PART III RESPONSES—STEPHEN M. GARDINER AND DAVID A. WEISBACH

DEBATING CLIMATE ETHICS

PART I

IN DEFENSE

OF CLIMATE ETHICS

STEPHEN M. GARDINER

1

How Will We Be Remembered?

Let us begin with a story[1]:

> Once upon a time, there was a generation that confronted great challenges and survived them. It struggled through a time of global financial collapse; defeated a frightening, destructive, and evil enemy; and ostensibly made the world safer for freedom and democracy for generations to come. This generation inherited a mess, but cleaned it up and passed on a better world to the future. It earned the moniker, "the most splendid generation."
>
> The most splendid generation was succeeded by another generation, "the bloopers." This generation had a reputation in its youth for grand visions and moral seriousness ("peace, love, and understanding"); however, when it actually came to hold the reins of power, it became consumed by the pleasures of the moment, and self-aggrandizement ("sex, drugs, and reality TV"). It paid scant attention to the concerns of the future, and indulged in whatever activities it could that brought soft comforts and profit in the short term, regardless of the long-term consequences. The bloopers deregulated financial markets, leaving the world vulnerable to a Great Depression–like crash; they provoked an international arms race and allowed the proliferation of weapons of mass destruction, making future wars more likely and more destructive; they polluted

the natural environment with wild abandon, undermining the future integrity of the world's climate system and food supply; and so on. In short, the blooper generation lived fast and loose, caring little whether others suffered greatly and died young as a result.

As things turned out, succeeding generations really did suffer serious harms (global financial collapses, horrific wars, environmental catastrophes, widespread famine, etc). Like the most splendid generation, they were left to clean up a mess.

This story has ethical import. The bloopers are a profligate generation. They squander the hard work of their predecessors, and inflict serious harms on their successors. Moreover, they do this mostly for the sake of cheap pleasures, and the comforts of easy living. Such a generation would receive harsh criticism from both the future and the past, and this criticism would be well deserved. They fail to discharge their intergenerational responsibilities. Too much goes wrong on their watch, and much of it is self-inflicted.

Sadly, the story has contemporary relevance. Many of us alive now, and especially those in the richer nations, are at risk of being remembered as members of a profligate generation—one that was recklessly wasteful, distracted, and self-absorbed. Moreover, our failures seem likely to be regarded especially harshly by the future, as they threaten to occur on a grand scale. The most serious involve an explosion in environmental degradation, with profound implications for all: globally, intergenerationally, and across species. If we do not address this issue, we may end up being remembered not just as a profligate generation, but as "the scum of the Earth," the generation that stood by as the world burned.[2]

It does not have to be this way. We are late, and dragging our feet. We have already taken greater risks than can plausibly be justified. However, there is still time, especially to head off the worst. If we can wake up to what we are doing and engage in meaningful action, we may still redeem ourselves. We can become the greenest generation yet. Given the scale of the challenge, that could make us the greatest generation of all.[3]

Notes

1. I thank audiences at the American Philosophical Association (Pacific Division), University of Graz, University of Leeds, University of Oregon, University of Victoria, and University of Washington. For comments, I am grateful to Richard Arneson, Michael Blake, Nir Eyal, Augustin Fragniere, Ben Gardiner, Avrum Hiller, Alex Lenferna, Marion Hourdequin, Lukas Meyer, Jay Odenbaugh, David Schlosberg, Dustin Schmidt, Ted Toadvine, and Allen Thompson. I also thank Dustin Schmidt and Alex Lenferna for their excellent assistance with referencing and copyediting. I am especially indebted to Kit Wellman, for all he is, and for steering this project home.
2. Stephen M. Gardiner, "Are We the Scum of the Earth?" in *Ethical Adaptation to Climate Change*, eds. Allen Thompson and Jeremy Bendik-Keymer (Boston: MIT Press, 2012), 241–260.
3. This prologue draws on Stephen M. Gardiner. *A Perfect Moral Storm* (Oxford: Oxford University Press, 2011), chapter 4.

2

Betraying the Future

We, the people, still believe that our obligations as Americans are not just to ourselves, but to all posterity.

We will respond to the threat of climate change, knowing that the failure to do so would betray our children and future generations.[1]

2.1 INTRODUCTION

In his second inaugural, President Barack Obama boldly asserted that climate change is an ethical issue, that our obligations to future generations are central to it, and that failure to meet those obligations would be a very serious moral matter (a "betrayal").[2] He is far from alone. Such pronouncements cross the political spectrum and other cultural divides, both nationally and internationally.

Ethical concerns are also central to the governing treaty for climate action, the United Nations' Framework Convention on Climate Change (UNFCCC), ratified in 1994 by all major countries, including the United States, China, the European Union, Russia, India, and Brazil. The UNFCCC states as its motivation the "protection of current and future generations of mankind," declares as its major objective the prevention of "dangerous anthropogenic interference" with the climate system, and requires that this be achieved while protecting ecological, subsistence, and economic values.[3] It also announces principles to guide the process that make heavy use of value-laden concepts, such as "equity," "common but differentiated responsibilities," the "special needs" of developing countries, and the "right" to development.

The thought that climate change is fundamentally an ethical issue is thus in many ways mainstream. Explicitly or implicitly, ethical concerns are widely held both to explain why we should be interested in the climate problem, and to structure acceptable responses. Nevertheless, such concerns have had very little influence over the making of more substantive international climate policy over the last twenty-five years. Instead, this has been dominated by short-term economic and strategic thinking.

This neglect arises in part because, in some circles, "ethics" is a "dirty" word, not to be mentioned in polite company unless to be ridiculed as obviously irrelevant, counterproductive, or even downright dangerous. Indeed, many in international relations and economics urge that ethics is best eliminated in global affairs quite generally, in favor of narrower considerations of national self-interest. Although (revealingly) this approach is applied only selectively to international issues, it has a strong influence on climate policy, especially in the United States. In particular, some (call them "the economic realists") insist that "pragmatically" harnessing national self-interest offers the *only* chance of success in combating the climate problem given the actual motivations of governments, since ethical concepts, and especially the key notion of justice, are hopelessly unfit for the purpose. This position fuels stark policy messages, such as Eric Posner's claim "you can have either climate justice or a climate treaty, not both,"[4] and perhaps the declaration of the US climate envoy, Todd Stern, to other negotiators, "If equity's in, we're out."[5]

My task in this volume is to defend ethics against such marginalization. For reasons that will become clear, I will not attempt to provide a comprehensive climate ethics. Instead my approach will be to explain why climate change is fundamentally an ethical issue, and why ethics is not easily eliminated from climate policy. My first chapter sketches the grounds for an ethical approach; the second argues against various versions of the anti-ethics position, including the politically influential version pressed by David Weisbach, and his colleagues Eric Posner and Cass Sunstein (the "Chicago lawyers"[6]); the third defends justice. Although the focus is climate, much of what I say applies to the role of ethics in international policy more generally.

My key claims will be as follows. First, *ethics gets the problem right*. Climate change is one instance of a distinctive challenge to ethical action: it is a perfect moral storm.[7] Moreover, ethical concerns (such as with justice, rights, political legitimacy, community, and humanity's relationship to nature) are at the heart of many of the decisions that need to be made. Consequently, climate policy that ignores ethics is at risk of "solving" the wrong problem.

Second, the *economic realists get the problem wrong, and dangerously so*. For one thing, they typically misdiagnose the climate problem as a traditional tragedy of the commons or prisoner's dilemma. Consequently, they miss central issues, and especially the critical intergenerational threat of what I call "a tyranny of the contemporary." Economic realism thus encourages "shadow solutions" that primarily serve the interests of affluent members of the current generation, including by creating illusions of real progress (e.g., Kyoto, Copenhagen).

Economic realists are also at risk of encouraging morally indecent policies, such as climate extortion. For instance, a key implication of the Chicago lawyers' "feasibility" approach is that the relatively poor, low-polluting nations who are the most vulnerable to climate impacts (e.g., Bangladesh) should "pay off" the (allegedly) less vulnerable large emitters (e.g., the United States, China) to stop polluting so heavily. Similarly, some "pragmatists" advocate passing the burdens of climate mitigation on to future generations through new forms of intergenerational debt.

Third, *the official rejection of ethics prevents us from raising central questions that need to be discussed.* In particular, although economic realists usually begin by insisting on the hegemony of narrow self-interest, they often end up appealing to wider ethical values, such as global welfare or limited intergenerational responsibility ("our children and grandchildren"). This vacillation not only renders such views unstable, but also undermines public deliberation. Though officially dismissed, ethics reemerges within a highly selective, morally loaded conception of self-interest left to be operationalized behind closed doors by policy professionals. Consequently, economic realism threatens a Trojan Horse. We, the people, are encouraged to quietly depart the scene, ceding power over the central ethical and geopolitical issue of our time to the "technocrats."

2.2 ETHICS FIRST

The third report of the UN's Intergovernmental Panel on Climate Change (IPCC) began by stating:

> Natural, technical, and social sciences can provide essential information and evidence needed for decisions on what

constitutes 'dangerous anthropogenic interference with the climate system.' At the same time, *such decisions are value judgments . . .*[8]

There are good grounds for this statement. Climate change is a complex problem raising issues across and between a large number of disciplines, including the physical and life sciences, political science, economics, and psychology, to name just a few. Still, without wishing for a moment to marginalize their contributions, ethics plays a fundamental role.

2.2.1 Evaluation

The first reason is that we *need ethical concepts to identify the relevant problem*. One issue is the challenge of the perfect moral storm (to which I return shortly). However, let me begin with a more general point. In stark physical terms, climate change (like most other things) might be seen as merely a series of events in the world. If we think that something should be done about it, it is because we *evaluate* those events, our role in bringing them about, and the alternatives. This evaluation gives us both an account of the problem, and constraints on what would count as relevant solutions. The critical question is what "values"[9] are in play when we do this.

Often, we do not notice that this question arises, since we assume that the relevant values are so widely shared that the answer is, or should be, "obvious" to everyone. Nevertheless, the values question is not trivial, since our answer shapes our whole approach. For example, when people say "murderers should be punished," we do not normally ask why; yet it makes a difference whether our reason is deterrence or retribution.

One way to highlight the values question in the climate context is to point out some highly restrictive ways of evaluating climate change. For instance, occasionally some with large fossil fuel holdings talk as if climate change is a "problem" *only* because tough emissions limits would make their assets worthless. For them, a good "solution" is to fund campaigns that question the science, and politicians who oppose action. So far, this "solution" has worked reasonably well in addressing their "problem." Nevertheless, theirs remains a poor description of what is really at stake in climate policy. One reason is that it is far too limited in what it takes into account; another is that these actors implicitly take their own narrow economic concerns as decisive over all other values.

A similar problem confronts the economic realists. Typically, they argue that the only thing that really matters to nation states as currently constructed is their short-term economic interests, usually understood in terms of domestic economic growth over the next couple of decades. However, on this account, it is not clear why climate change is much of a problem. Given the long time lags involved, most climate impacts, including many of the most serious, are not short-term in this sense, nor narrowly economic. Moreover, those that will occur in the next few decades are likely already "in the cards," in the sense that we are already committed to them, either by past emissions or by those that are now, practically speaking, inevitable. Consequently, a policy that tried to address climate change with an exclusively "next decade or two" focus would confront only a very small set of the relevant impacts, and probably miss the most important (e.g., the potentially catastrophic). Such a policy would probably also employ the wrong methods. For example, if all that mattered were domestic economic impacts for the next twenty years, but these were more or less "in the cards," mitigation would likely seem pointless, or even counterproductive. From the "few decades" point of view, it seems much better to put the resources into offsetting the immediate threats (e.g., through national adaptation). Why not, if the main point of mitigation would be to reduce later effects that fall mainly elsewhere and on others?[10]

In my view, better explanations of the climate "problem" appeal to a much more extensive, but also widely shared, set of values. The climate problem that *should* concern public policy is global, intergenerational, and ecological in scope, and the most important concerns are ethical, including considerations of justice, rights, welfare, virtue, political legitimacy, community, and our relationship to nature. If public policy neglects such concerns, its account of the problem to be solved is impoverished, and the associated solutions quickly become grossly inadequate. For example, we do not "solve" the climate problem if we inflict catastrophe on future people, or devastate poor African nations, or rapidly accelerate the pace of mass extinction. We can summarize this point by saying that alleged solutions face a set of *intelligibility constraints*. Economic realists (and others) must explain what problems their "pragmatic" policies seek to solve, and why these are the most pressing.

Importantly, there are signs that some intelligibility constraints are already beginning to bite. For instance, some world leaders criticized the Copenhagen Accord's proposal to interpret "dangerous climate change" as that which exceeds a two-degree limit. Mohamed Nasheed, President of the Maldives, complained:

> Anything above 1.5 degrees, the Maldives and many small islands and low-lying islands would vanish. It is for this reason that we tried very hard during the course of the last two days to have 1.5 degrees in the document. I am so sorry that this was blatantly obstructed by big-emitting countries.

More dramatically, Lumumba Stanislaus Di-aping, lead negotiator of the G-77 group of developing countries, protested:

> [The draft text] asks Africa to sign a suicide pact, an incineration pact in order to maintain the economic dominance of a few countries. It is a solution based on values, the very same values in our opinion that funneled six million people in Europe into furnaces.[11]

Whatever one thinks of the rhetoric, the ethical worry is clear. Without justice to developing nations, what (or who's) problem does a climate treaty solve?

Elsewhere I argue that the dominance of short-term and narrowly economic concerns favored by economic realists goes a long way towards explaining the past failures of international climate policy in Kyoto and Copenhagen. Although these were disastrous in ethical terms, arguably they were great "successes" in achieving the modest ambitions of the current generation of the most powerful. Most notably, for many they perpetuated a "dangerous illusion" of progress that facilitated an ongoing strategy of distraction and delay.[12] As we shall see, such "shadow solutions"—reflecting only the limited concerns of those with the power to act rather than the central ethical concerns—are persistent threats in the climate case.

2.2.2 Policy Questions

The second reason that ethics plays a fundamental role in climate change is that *ethical considerations are right at the heart of the main policy decisions that must be made*, such as how quickly to reduce greenhouse gas emissions over time, how those emissions that are allowable at a given time should be distributed, and what should be done to address unavoided impacts.

Suppose, for instance, one were deciding where to set a global ceiling on emissions for a particular time. This decision depends in large part on how the interests of the current generation are weighed against those of future generations. At one extreme, giving absolute priority to the interests of the future probably means ceasing emissions very quickly, even if this involves severe sacrifices for the current generation; at the other extreme, continuing high levels of emissions—as we are currently doing—suggests giving the future no weight at all. Presumably, neither extreme is justified, but determining precisely where the appropriate balance lies is an ethical question (and a difficult one).

Similarly, ethical questions pervade the issue of how to distribute emissions under a ceiling. Distributive decisions depend in part on background beliefs about the appropriate role of energy consumption in people's lives, the relevance of historical responsibility, and the current needs and future aspirations of particular societies. For instance, should those in severe poverty get greater access than the affluent, or do those who have already invested in fossil-fuel intensive infrastructure have a prior claim? Again, the ethical questions are serious and central.

2.2.3 Ethical Challenge

The third reason that ethics is fundamental is that climate change *presents a severe ethical challenge*. It throws down the gauntlet to us as ethical agents, and especially to our moral and political systems. Specifically, climate change is an early instance of a problem that poses a profound ethical test for humanity and its institutions. I call this problem, "the perfect moral storm." The ongoing political inertia surrounding climate action suggests that so far we are failing that test.

Let us say that a perfect storm is an event constituted by an unusual convergence of independently harmful factors where this convergence is likely to result in substantial, and possibly catastrophic, negative outcomes. The phase "perfect storm" became prominent in popular culture through Sebastian Junger's book and Wolfgang Peterson's subsequent movie starring George Clooney.[13] Junger's tale is based on the true story of the *Andrea Gail*, a fishing vessel caught at sea during a convergence of several independently powerful storms. The sense of the analogy is that climate change is a perfect *moral* storm because it involves the convergence of a number of factors that threaten our ability to behave ethically.

As climate change is a complex phenomenon, I cannot hope to identify all of the ways in which its features create challenges for ethical behavior. Instead, I will highlight four especially salient threats—analogous to the storms that hit the *Andrea Gail*—that converge in the climate case. These "storms" arise in the global, intergenerational, ecological, and theoretical dimensions. Each is serious in its own right. However, their interaction also helps to exacerbate a lurking problem of moral corruption that may be of greater practical importance than any one storm considered in isolation.

2.3 THE GLOBAL STORM

In the policy world, the climate challenge is usually understood in spatial, and especially geopolitical, terms.

2.3.1 The Basic Global Storm

We can make sense of this by focusing on three important characteristics. The first is the spatial *dispersion of causes and effects*. Climate change is a truly global phenomenon. Emissions of greenhouse gases from any geographical location on the Earth's surface enter the atmosphere and then play a role in affecting climate globally. Hence, their effects are not realized solely at their source, either individual or geographical, but rather are dispersed across all regions of the Earth.

The second characteristic is *fragmentation of agency*. Climate change is not caused by a single agent, but by a vast number of individuals and institutions (including economic, political, and social institutions) not fully unified by any comprehensive structure of agency. This poses a challenge to humanity's ability to respond.

In the spatial dimension, fragmentation of agency is usually understood as arising out of the shape of the current global system, dominated by nation states, and in terms of the familiar theoretical model of the prisoner's dilemma, or what Garrett Hardin calls a "tragedy of the commons."[14] Weisbach and his colleagues also endorse this approach.[15]

Later I will argue that the standard model is a *dangerous misdiagnosis* that threatens good policy. However, first let us explain it. The relevance of the prisoner's dilemma scenario is easiest to show by focusing on a paradigm case: overpollution. Suppose that a number of distinct agents are trying to decide whether or not to engage in a polluting activity. Assume for the moment that each is concerned only with its own interests, narrowly construed, and that the situation can characterized as follows:

(PD1)   It is *collectively rational* to cooperate and restrict overall pollution: each agent prefers the outcome produced by everyone restricting its individual pollution over the outcome produced by no one doing so.

(PD2)   It is *individually rational* not to restrict one's own pollution: when each agent has the power to decide whether or not it will restrict its pollution, each (rationally) prefers not to do so, whatever the others do.

Agents in such a situation find themselves in a paradoxical position. On the one hand, given (PD1), they understand that it would be better for everyone if every agent cooperated; but, on the other hand, given (PD2), they also all prefer to defect. Paradoxically, then, if all individual agents act rationally in terms of their own interests, then they collectively undermine those interests.

For current purposes, assume that a tragedy of the commons is roughly a prisoner's dilemma involving a common resource.[16] This has become the standard analytical model for understanding large-scale environmental problems, and climate change is no exception. Typically, the reasoning goes as follows. Conceive of climate change as an international problem where the relevant parties are individual countries, who represent the interests of their populations in perpetuity. Then (PD1) and (PD2) appear to hold. Individual states accept that allowing climate change to continue unabated is bad for them, that cooperation is needed to address it, and that it would be in their interests for all to cooperate (i.e., (PD1)). However, each state also believes that when it comes to making its own decisions about what to do, it is better not to cooperate, since this choice is better on strategic grounds (i.e., (PD2)). Specifically, on the one hand, if others cooperate, it is better to defect, since then one can receive the benefits of cooperation (i.e., meaningful reductions in overall climate risk) without having to pay any of the costs; however, on the other hand, if others fail to cooperate, it is also better not to cooperate, since otherwise one would pay costs without receiving the benefits (e.g., since meaningful reductions require cooperation). Unfortunately, this pattern of reasoning leads to a paradoxical result: if each country reasons in the same way, no one cooperates, and each ends up worse off by its own lights than they would if all cooperated. This result is aptly termed a tragedy: the problem seems self-inflicted and the behavior self-destructive.

If climate change is a normal tragedy of the commons, this is challenging, but also encouraging. Given (PD1), there is a sense in which each country should be motivated to find a way out of the paradox, so that all can secure the better, cooperative outcome. Moreover, in the real world, commons problems are often resolvable under certain circumstances, and climate change seems to fit these desiderata.[17] In particular, commons problems are often resolved if the parties benefit from a wider context of interaction; and this appears to be the case with climate, since countries cooperate on a number of broader issues, such as trade and security.

Unfortunately, this brings us to the third characteristic of the basic global storm, *institutional inadequacy*. The usual means for resolving commons problems under favorable conditions is for the parties to agree to change the existing incentive structure by introducing a system of enforceable sanctions. (Hardin memorably labels this "mutual coercion, mutually agreed upon.")[18] Such a system transforms the decision situation by foreclosing the option of free riding, so that the collectively rational action also becomes individually rational. Theoretically, then, matters seem simple. Nevertheless, in practice the need for enforceable sanctions poses a challenge at the global level because of the limits of our current, largely national, institutions, and the lack of an effective system of global governance. In essence, global regulation of greenhouse gas emissions seems required, including a reliable enforcement mechanism; however, the current global system—or lack of it—renders this difficult, if not impossible.

The standard (spatial) analysis thus suggests that the main thing needed to address climate change is an effective system of global governance (at least for this issue). This is a tall order; still, there is a sense in which it remains good news. In principle at least, it ought to be possible to motivate countries to establish such a regime, since they should recognize that it is in their best interests to eliminate free riding and so make genuine cooperation the rational strategy at the individual (i.e., national) as well as collective (i.e., global) level.

2.3.2 Exacerbating Factors

Alas, other features of the climate case make the necessary global agreement more difficult, and so exacerbate the basic global storm.

The first is *uncertainty* about the precise magnitude and distribution of climate impacts. Lack of trustworthy data about national costs and benefits casts doubt on the truth of (PD1): the claim that each country prefers the outcome produced by everyone restricting pollution. Perhaps, some wonder, we might be better off with at least some climate change. More importantly, some (e.g., the United States) might ask whether they will at least be relatively better off under climate change than others (e.g., Bangladesh), and so might get away with paying less to avoid the costs of cleaning up. (Such considerations are emphasized by the Chicago lawyers, and fundamental to their analysis.)

In other contexts, uncertainty might not be so important. However, the second exacerbating feature of the climate problem is its *deep roots* in the infrastructure of many current civilizations. Carbon dioxide emissions are predominately brought about by the burning of fossil fuels for energy, and this energy supports most existing economies. Given that deep cuts are needed over time, such actions are likely to have profound implications for the basic economic organization of developed countries and the aspirations of others. One implication is that those with vested interests in the continuation of the current system—e.g., many of those with substantial political and economic power—will resist such action. Another is that, unless ready substitutes are found, substantial mitigation can be expected to have considerable repercussions for how humans live and how societies evolve. In short, climate action is likely to raise serious, and perhaps uncomfortable, questions about who we are and what we want to be.[19] Among other things, this suggests a *status quo* bias in the face of uncertainty. Contemplating change is often uncomfortable; facing fundamental change may be unnerving, even distressing. Since the social ramifications of action appear to be large, perspicuous, and concrete, but those of inaction appear uncertain, elusive and indeterminate, it is easy to see why uncertainty might exacerbate social inertia.[20] If we already dread doing something, even weak reasons not to can seem especially tempting.

The third feature of climate change that exacerbates the global storm is *skewed vulnerabilities*. Climate change interacts in unfortunate ways with the present global power structure. For one thing, responsibility for historical and current emissions lies predominantly with the richer, more powerful nations, and the poor nations are badly situated to hold them accountable. For another, the limited evidence on regional impacts suggests that poorer nations are most vulnerable to the worst. Finally, recognizing and acting on climate creates a moral risk for the richer nations. Implicitly, it embodies an acknowledgement that there are international norms of ethics and responsibility, and reinforces the idea that international cooperation on issues involving such norms is both possible and necessary. Hence, it may encourage attention to other moral defects of the current global system, such as global poverty, human rights violations, economic injustice, the legacy of colonialism, and so on. If richer nations are not ready to engage on such topics, this gives them further reason to avoid serious climate action.

2.3.3 Rejecting the Traditional Model

The claim that climate change is a tragedy of the commons (or prisoner's dilemma) played out between nation states is pervasive in academia, policy, and the popular press. Nevertheless, I believe it involves a dangerous misdiagnosis.

Some reject the model as too pessimistic, claiming that game theory relies on false and morally problematic assumptions. Actual agents, they say, (whether people or nations) are not narrowly self-interested; and such claims have the effect of ruling out more generous motivations—and the possibility of ethical action—by definition. Consequently, the traditional model is not an appropriate guide to policy.

I agree that the usual motivational assumptions should be resisted, but not that this justifies rejecting game-theoretic methods. My first reason is that the basic approach is compatible with more generous motivational assumptions. For instance, agents often value others deeply, but nonetheless primarily in a self-referential or time-indexed way. For example, they value *their own* children or community, or *their own* time and place. Game theory helps to reveal how this can also result in tragedy (e.g., when anxious parents competitively "hothouse" their children).

My second reason is that game-theoretic models are often a good guide to what to expect when wider values are present but not adequately reflected in existing institutions, so that they play little role in how decisions are actually made. In such situations, the result picked out by narrow, self-interested, self-referential, or time-indexed motivations is often the natural "default." This does not show that people lack wider values, only that there are no effective channels to register them, or at least that the channels available to narrower values are much more effective. Importantly, this seems plausible in the climate case, since markets and standard election cycles provide routes for short-term economic concerns to drive decision making that seem lacking for global, intergenerational, and ecological concerns. Game theoretic models thus help us to foresee the consequences of institutional failure.

Nevertheless, I maintain that we should reject the standard tragedy of the commons analysis of climate change. This is not because it is too pessimistic, but rather because it is *overly optimistic*. First, it obscures basic issues of international fairness. For instance, in framing the policy challenge as one of securing cooperation between parties through a process of self-interested bargaining, the standard model marginalizes important ethical issues, such as skewed vulnerabilities and background injustice. Such factors allow powerful bargainers to take advantage of the circumstances and create compound injustices (cf. the complaints of the Maldives and G77). Ignoring them leads policymakers to underestimate the real challenges they face.

Second, and even more importantly, the tragedy of the commons analysis ignores vital dimensions of the climate challenge, and in ways that threaten the integrity of the whole global storm approach. Most notably, the standard model conceives of climate change as essentially an international problem, under the assumption that the relevant parties are individual countries *who effectively represent the interests of their populations in perpetuity*. This assumption is dubious in general. As we shall see, it is especially reckless for climate change.

2.4 THE INTERGENERATIONAL STORM

The spatial characteristics of the global storm (dispersion of causes and effects, fragmentation of agency, and institutional inadequacy) can also be seen from a temporal perspective.

2.4.1 The Basic Intergenerational Storm

Consider first the temporal dispersion of causes and effects. Human-induced climate change is a severely lagged phenomenon. This is partly because some of the basic mechanisms set in motion by the greenhouse effect—such as sea-level rise—take a very long time to be fully realized, and partly because, once emitted, molecules of the most important anthropogenic greenhouse gas, carbon dioxide, spend a surprisingly long time in the atmosphere where they continue to have climatic effects.[21]

Let us dwell for a moment on the second factor. In its early reports, the IPCC reported that the average time spent by a molecule of carbon dioxide in the atmosphere is in the region of five to two hundred years.[22] This estimate is long enough to suggest a serious lagging effect; nevertheless, it obscures the significant percentage that remains for much longer. The latest IPCC report highlights that, depending on how much is produced, "about 15%–40% of CO_2 emitted until 2100 will remain in the atmosphere longer than a thousand years," and 10%–25% after 10,000 years.[23]

Temporal lagging has at least three important implications. First, climate change is *resilient*, in the sense of not being easily countered or reversed. For instance, the process cannot be quickly undone simply by reducing new emissions. Achieving goals such as restoring or retaining a familiar climate requires advance planning.

Second, climate change is *significantly backloaded*: the full effects of previous emissions have yet to be realized. For example, according to the IPCC, by 2000, in addition to the then-observed rise in global average temperature of 0.6°C, we had already effectively committed the planet to at least another 0.6°C by 2100.[24] Similarly, some say that the collapse of the West Antarctic Ice Sheet is now inevitable and irreversible, although the process will take many centuries to play out. We have not yet seen all that we have done.

Third, the flip side of backloading is that climate change is also *substantially deferred*: the full effects of ongoing emissions will not be realized until well into the future. Our choices will continue to have profound effects long after we are gone.

Temporal dispersion creates a number of problems. First, *resilience* implies that delays in action have serious repercussions for our ability to manage the problem. Second, *backloading* brings on serious epistemic difficulties, especially for normal political actors. For instance, it makes it hard to grasp the connection between causes and effects, and this may undermine the motivation to act; backloading also suggests that by the time we realize that things are bad, we may already be locked in to suffering much more change, undermining the ability to respond. Third, *substantial deferral* calls into question the ability of standard institutions to address the problem. For instance, democratic political institutions have relatively short time horizons—the next election cycle, a politician's political career—making it doubtful whether they have the wherewithal to deal with temporally distant impacts. Even more seriously, substantial deferral is likely to undermine the will to act. This is because there is an incentive problem: the bad effects of our current emissions are likely to fall, or fall disproportionately, on future generations, whereas the benefits of emissions accrue largely to the present.

The incentive problem is especially dangerous given the second feature of the intergenerational storm: temporal fragmentation of agency. Those with power and those affected are spread across a very long period of time. This threatens to bring on a new kind of collective action problem: a tyranny of the contemporary.[25] To illustrate it, let us first jettison the standard assumption (e.g., in the tragedy of the commons analysis) that governments reliably represent the interests of both their present and future citizens. For instance, suppose instead that they are biased towards the shorter-term concerns of the current generation. Then since the benefits of carbon emissions are felt primarily by current people (in the form of cheap energy), whereas the costs (in the form of the risk of severe climate change) are substantially deferred to future generations, the current generation faces a temptation to "live large" and pass the bill on to the future. Worse, the temptation is iterated. As each new generation gains the power to decide whether or not to act, it has the same opportunity to pass the buck. If multiple generations succumb to this temptation, we might expect an accumulation of effects in the future, rapidly increasing the risks of severe or catastrophic impacts. Iteration breeds escalation.

Note that, as with the tragedy of the commons, although the most obvious driver of a tyranny of the contemporary is each generation ruthlessly pursuing its own self-interest (narrowly defined), tyranny can arise for other reasons. For instance, perhaps people are very altruistic within their own generation, but indifferent more broadly; or perhaps their institutions disproportionately register shorter-term interests, so that their (genuine) intergenerational concerns are crowded out; or perhaps a given culture is shallow and apathetic (rather than ruthless), and so unthinkingly adopts a parochial form of life that is good neither for itself nor for others. All that is required for a tyranny of the contemporary is that the concerns that turn out to be most effective are (for whatever reason) dominantly *generation-relative*.

The deep threat posed by the tyranny of the contemporary is easiest to see if we compare it to the traditional prisoner's dilemma. Suppose we consider a pure, and in many ways optimistic form of tyranny,[26] where generations do not overlap and (for whatever reason) each is concerned only with what happens during the timeframe of its own existence. Suppose then that the situation can be (roughly) characterized as follows:

> (PIP1) It is *collectively rational* for most generations to cooperate: (almost) every generation prefers the outcome produced by everyone restricting pollution over the outcome produced by everyone overpolluting.

> (PIP2) It is *individually rational* for all generations not to cooperate: when each generation has the power to decide whether or not it will overpollute, each generation (rationally) prefers to overpollute, whatever the others do.

Call this the "Pure Intergenerational Problem" (PIP).

The PIP is worse than the prisoner's dilemma. On the one hand, (PIP1) is worse than (PD1) because not everyone prefers the cooperative outcome. The first generation benefits neither from the cooperation of its successors, nor from holding back on its own pollution. Worse, this problem becomes iterated. Since subsequent generations have no reason to comply if their predecessors do not, noncompliance by the first generation has a domino effect that undermines the collective project. On the other hand, (PIP2) is worse than (PD2) because the reason for it is deeper. Both claims hold because the parties lack access to mechanisms (such as enforceable sanctions) that would make defection irrational. However, whereas in normal prisoner's-dilemma-type cases this obstacle is largely practical, and can be resolved by creating appropriate institutions, in the PIP it arises because the parties do not coexist, and so (on the face of it) seem unable to influence each others' behavior through the creation of appropriate coercive institutions. In addition, the standard solutions to the prisoner's dilemma are unavailable: one cannot appeal to a wider context of mutually-beneficial interaction, nor to the usual notions of reciprocity.

The PIP is a hard problem. Fortunately, the real world deviates from it: generations do overlap, and many people within each generation do care about what happens to other generations. Still, it is not clear that these facts alone are sufficient to diffuse the challenge. For instance, perhaps overlap does not make the right kind of difference or make it at the right time (e.g., we care about our grandchildren, but only when we are old enough to see them grow); or perhaps intergenerational concern exists but is substantially weaker than generation-relative interests (e.g., we are interested in humanity's survival, but not as much as in our Hummers). Given this, we must still address other versions of the tyranny of the contemporary, including degenerate versions of the PIP. In my view, an especially important form arises when *contemporary institutions are structured so that generation-relative interests dominate concern for the future*. Call this, "the Generation-Relative Institutional Problem" (GRIP).[27] Even otherwise ethical generations—with strong concern for future people—may be defeated by the GRIP.

The upshot of this section is that, when instantiated, paradigm tyranny of the contemporary models (such as the PIP and GRIP) are less optimistic than the mainstream prisoner's dilemma analysis of climate change. For instance, on some versions of the intergenerational problem, current populations may not be motivated to establish a fully adequate global regime, since, given the temporal dispersion of effects—and especially backloading and deferral—such a regime may be neither in *their* interests, narrowly conceived, nor respond to their generation-relative concerns. Alternatively, on other versions, they may be appropriately motivated, but remain stuck in the GRIP, lacking the institutional wherewithal to make intergenerational concern count in the world of policy. In my view, these are especially serious worries in the climate case, since the intergenerational dimension dominates.

2.4.2 Exacerbating Factors

The threat of a tyranny of the contemporary is bad enough considered in isolation. Unfortunately, in the climate context, it is also subject to morally relevant multiplier effects.

First, the threat of escalation is especially serious. In failing to act appropriately, the current generation is not simply passing an existing problem along to future people, it adds to it, making the problem worse. In the last fifty years, global emissions have increased more than 400%. Moreover, the annual growth rate has increased from around 1% in the 1990s to closer to 3% more recently. (Notably, given compounding, even 2% yields a 22% increase in emission rates in 10 years and 35% in 15 years.[28]) In addition, inaction increases future mitigation costs, by allowing additional long-term investment in fossil fuel-based infrastructure, especially in poorer countries (e.g., estimates suggest a ten-year delay could increase costs by 40%).

Second, insufficient action may make some generations suffer unnecessarily. Suppose that, at this point in time, climate change seriously affects the prospects of generations A, B, and C. If A refuses to act, this may make the effects continue for longer, harming D and E. Some take this to violate a fundamental moral principle of "Do No Harm."[29]

A third multiplier is that A's inaction may create situations where *tragic choices* become necessary. For instance, one generation may create a future where its successors would be *morally justified* in making other generations suffer unnecessarily. Suppose, for example, that A could and should act now in order to limit climate change such that D would be kept below some crucial climate threshold. If passing the threshold imposes severe costs on D, D's situation may be so dire that it is morally justified in take action that will severely harm generation F, such as emitting even more carbon. For instance, under some circumstances, normally impermissible actions that harm innocent others may become morally permissible on grounds of self-defense.

Notably, one way in which A might behave badly is by creating a situation such that D is forced to call on the self-defense exception, and thereby inflict severe suffering on F. This is a morally serious matter for both D and F. For one thing, even if D is in some sense justified in harming F, taking this action may still *mar* their own lives from the moral point of view, and in a way that is difficult to live with, as in a Sophie's choice situation. (Who wants to be responsible for inflicting such suffering?) For another thing, in a tyranny of the contemporary, the transmission of such tragic choices can become iterated: perhaps F must also call on the self-defense exception, and so inflict harm on generation H, and so on. If so, A may ultimately be responsible for initiating a whole chain of suffering that mars the lives of many of its successors. Arguably, this is an especially grievous kind of moral wrong, which profoundly affects ethical evaluation of A, giving it even stronger reasons to desist.[30]

2.4.3 Revisiting the Global Storm

Interestingly, the tyranny of the contemporary analysis of the intergenerational storm further undermines the standard (tragedy of the commons) account of the global storm.

On the one hand, if generational ruthlessness dominates, the global collective action problem of most interest to existing institutions might be that of finding ways for the current generation in powerful nations to cooperate in passing costs onto the future. This "problem"—how to facilitate a tyranny of the contemporary—looks easier to resolve than a standard tragedy of the commons (e.g., governments committed to intergenerational buck-passing may have no reason to defect). Yet solving it is not a task for an ethical public policy: it constitutes a shadow solution to climate change.

On the other hand, if, as I have suggested, the main driver of the tyranny of the contemporary is institutional inadequacy, the necessary new institutions might radically change the shape of the global collective action, rendering it much less tragic. For instance, institutions that successfully embody strong intergenerational concern will presumably see climate change as a serious threat, and so not even be tempted to defect from robust action. Consequently, their existence would radically transform the spatial dimension of climate policy.

2.5 THE ECOLOGICAL STORM

The third storm is ecological.

2.5.1 The Basic Ecological Storm

Once again, there is *dispersion of causes and effects*; this time across species.[31] Anthropogenic climate change has profound implications for other animals, plants and eco-systems. Often, this encourages a distinct form of ecological buck-passing. Consider the simple metaphor of *Kick the Dog*. In the (terrible) old children's story, the farmer kicks his wife, his wife kicks the child, and the child kicks the dog. In the climate case, the parallel is likely to be that the current rich "kick" the poor, both "kick" future generations, and all pass the "kicking" on to other species through the ecological systems on which they depend. In other words, the initial bad behavior sets off a chain reaction towards the end of which stand not just the most vulnerable humans, but also many animals, plants, and places (e.g., the polar bear, big cats, rainforests, the Arctic). Moreover, if and when the natural world kicks back, it is likely to induce further cycles of buck-passing.

There are many reasons to decry this scenario. The first is narrowly anthropocentric: the instrumental value of non-human nature to humans, in terms of things like biological resources and "ecosystem services." Although compelling insofar as it goes, this reason remains limited. For example, years ago I came across a magazine article by an economist that argued that climate change is not a problem because future humans could build massive domes on the earth's surface to live in. On the instrumental view, *Dome World* is problematic only because it is not currently technically feasible, or else too expensive by comparison to protecting the natural world. Humanity's problem is that we do not yet have the means to trash the Earth and move on, as they do in the film *Wall-E*. (Trashing is off the table only because we currently lack a good escape route.)

A second, but broader, anthropocentric view denies this. It claims that nonhuman nature also plays a constitutive role in flourishing human lives in a way that cannot be replicated in *Dome World* even if such a world were possible. For example, life on the spaceships of *Wall-E*, watching TV and drinking soda, is just too shallow to be really good for human beings.

Some accept a modest version of the constitutive view, believing that it holds under current circumstances, but that technology will ultimately provide good substitutes for the natural world (e.g., virtual reality devices like Star Trek's "holodeck"). My own view is that we should look deeper. For one thing, nonhuman nature represents the world from which we evolved, and against which we understand and define ourselves. Our relationship with it plays a considerable role in determining our self-conception. Trashing it therefore harms us in a deeper way than concern for ecosystem services suggests. In some ways it is an attack on who we are and can be, and one proceeding from a very narrow conception of the human good. To that extent, easy talk of substitutes seems a little like the view that the Mona Lisa can eventually be replaced, without loss, by an image on a high definition TV. Although we can see why someone might take this view, and even be unsure how to combat it, it is significantly out of step with what most of us believe.

A third view is straightforwardly nonanthropocentric: at least some nonhuman entities, relationships, and processes have value beyond their contribution to human projects.[32] Such positions are often treated as marginal, as well as too controversial to be taken seriously in both public policy and mainstream political philosophy. Still, I have come to believe that this is a mistake. In fact, most of us are nonanthropocentrists, to at least some extent. We believe (sometimes implicitly) that at least some nonhuman entities warrant at least some ethical consideration apart from their contribution to human projects. For example, it is very common to think that entities such as the great apes, polar bears, giant sequoias, and the Great Barrier Reef deserve protection and respect even if we never visit them, benefit from them, and so on. Indeed, this is a central reason why *Kick the Dog* and *Dome World* bother us. Yet if this is our view, then neither public policy nor philosophy should ignore it. To treat such beliefs as akin to mere personal preferences, or (worse) as not even deserving the political standing of such preferences, is to dismiss them too readily and without regard for their content, which fundamentally involves judgments about what really matters, not mere questions of taste.[33]

Like the previous storms, the ecological storm may also involve *fragmentation of agency*, albeit this time in an extended sense. We tend to see agency strictly speaking as an exclusively human preserve. However, to describe the rest of nature as merely a passive victim of human action would be deeply misleading: the nonhuman world behaves in ways that are independent of humans and reflect (at least some) nonhuman concerns. Moreover, traditionally, many environmentalists have seen this *independence* of nature from human purposes as an important source of its value, and so have aspired to respect and accommodate that independence to at least some extent.

This suggests that *institutional inadequacy* is also an issue in the ecological storm. Current institutions remain largely blind to, or biased against, the whole issue of accommodation. Furthermore, while any attempt at accommodation would already involve reconciling a vast number of different aims (human and nonhuman), and integrating them into a broad vision of flourishing (for them and for us), this challenge becomes even more daunting in a world of anthropogenic climate change.

2.5.2 Exacerbating Factors

Several considerations exacerbate the ecological storm in the climate case. The first is the deep and pervasive reach of human influence. Serious anthropogenic climate change looks likely to shape the basic conditions of life on the planet for the nonhuman world, profoundly affecting the lives, roles, and very existence of countless other creatures and ecosystems. It is one of the main drivers of what many now call the *Anthropocene*, a new geological era defined by human domination. This provides a large challenge for how to understand "independence" and "accommodation." For instance, some claim that the onset of the Anthropocene implies that wild, independent nature will soon no longer exist. One issue is that, if this is a distinct source of value and climate change extinguishes it from the Earth, the ethical stakes are high. Another is what to say about the value of the nonhuman world that remains.

A second exacerbating factor is the radical expansion of human responsibility. The pervasive influence of humanity implies that "what happens" in nature is fast becoming *what we do*. If "with great power comes great responsibility," the implications seem dramatic. Just as formerly "natural" disasters (such as superstorms, profound drought, flooding, etc.) can increasingly be seen as human-made disasters, so we are at risk of becoming increasingly implicated in the nonhuman world, including in its less savory aspects (such as extinction, predation, disease, and so on). For instance, if in the future polar bears go extinct through the lack of summer sea ice, or pine beetles shift their range southwards destroying the redwoods, these will be things that we have facilitated, and for which we share responsibility. Nature, previously acknowledged to be "red in tooth and claw" but no one's fault, suddenly belongs to us.[34]

A third exacerbating factor is the possibility that this situation encourages a push toward further human *colonization* of nature. For instance, if the nonhuman world is seen as having become thoroughly our responsibility, some will claim that it is now our job to clean it up, morally speaking (e.g., by "policing" interactions among nonhumans, to create a "sanitized" nature). While some view this as an obvious next step, others are horrified. Again, climate change amplifies serious questions about the human relationship to the nonhuman world.

2.6 THE THEORETICAL STORM

The fourth storm is theoretical: we are extremely ill equipped to deal with many problems characteristic of the long-term future. In my view, even our best theories—whether economic, political, or moral—struggle to address basic issues such as intergenerational equity, contingent persons, nonhuman animals and nature. Climate change involves all these and more. Given this, humanity appears to be charging into an area where we are theoretically inept, in the (nonpejorative) sense of being unsuited for (e.g., poorly adapted to), or lacking the basic skills and competence to complete, the task.

This criticism applies directly, and particularly seriously, to the dominant approach in public policy today (and incidentally the favorite of the Chicago lawyers): standard economic cost-benefit analysis (Market CBA). I develop this criticism in chapter 3. Mirroring economic realist complaints about ethics, my claim will be that Market CBA lacks the tools for the job at least without ethical underpinnings. Consequently, particular cost-benefit analyses tend to hide their own ethical assumptions from view, rather than making them explicit and thereby subject to scrutiny. As a result, such analyses often risk violating basic ethical constraints.

These concerns suggest that taking ethics seriously is essential to resolving the theoretical storm. Still, mine is not a partisan claim. In my view, the theoretical storm also afflicts major theories in moral and political philosophy, such as utilitarianism, Rawlsian liberalism, human rights theory, libertarianism, virtue ethics, and so on. I conceive of such theories as major research programs that evolve over time (rather than, say, as sets of propositions). Whatever their other merits, in their current forms these research programs appear to lack the resources needed to deal with problems like climate change. Moreover, it seems likely that in evolving to meet them, they will be substantially, and perhaps radically, transformed.[35]

Importantly, I am not claiming that *in principle* such theories have nothing to say. On the contrary, at a superficial level, it is relatively easy for the standard research programs to assert that their favored values are relevant to climate change and license condemnation of political inertia. Severe and catastrophic forms of climate change pose a big enough threat that it is plausible to claim that most important values (e.g., happiness, human rights, freedom, property rights, etc.) are threatened. Surely, then, proponents of such values will think that something should be done. Nevertheless, in my view the more important questions are how precisely to understand the threat, and what should be done to address it. On such topics, the standard research programs seem curiously oblivious, complacent, opaque, even evasive. In particular, so far they have offered little guidance on the central question of the kinds of institutions that are needed to confront the problem, and the specific norms that should govern those institutions. Though this situation is starting to change as theoretical attention shifts (after twenty-five years of waiting), there is still a very long way to go.

Of course, none of this implies that ethics has nothing to offer even in these early days. As we wait for suitably robust theories—whether economic, political, or moral—to emerge, moral and political philosophy can be useful in guiding an ethics of the transition. For instance, concepts such as justice can put limits on how we should think about the problem, even if those limits are not yet fully articulated. (As Amartya Sen once put it, sometimes "it is better to be vaguely right than precisely wrong.") They are also seeds from which fuller theories (including more ideal theories) can grow, and their historical development in other contexts can provide meaningful guidelines. Nevertheless, we should admit from the outset that we are not there yet, and much work remains to be done. Sometimes an ethics of the transition must take on the task of guiding us forward even when we are not sure precisely where to go.

2.7 MORAL CORRUPTION

One reason that an ethics of the transition is urgently needed is that the convergence of the four storms creates a further threat, this time to the way we think and talk about climate change. Faced with the temptations of the global, intergenerational and ecological storms, and the cover provided by the theoretical storm, it is easy for our reasoning to become distorted and perverted. Public discourse in particular is under threat.[36] As Robert Samuelson puts it in another intergenerational context, "There's a quiet clamor for hypocrisy and deception; and pragmatic politicians respond with . . . schemes that seem to promise something for nothing. Please, spare us the truth."[37] Given this, there is a role for "defensive ethics" that combats such tendencies.

Corruption of the ways in which we think and talk can be facilitated in a number of ways, including distraction, complacency, unreasonable doubt, selective attention, delusion, pandering, false witness, and hypocrisy. Merely listing such strategies is probably sufficient to make my main point. (Close observers of climate politics will recognize many of them.) So, here let me offer just two illustrations.

The first concerns unreasonable doubt. At the time of writing, the basic science underlying concern about climate change is the subject of a widespread, enduring and strengthening consensus that has been repeatedly stressed in international reports, and endorsed by the national academies of major nations.[38] Nevertheless, in the political and broader public realms, the level of doubt, distrust, and even outright hostility to this consensus has, if anything, increased as scientific understanding has developed. It is often (rightly) said that this is partly due to campaigns of disinformation, and a widespread misunderstanding within the media and broader public of the role of legitimate skepticism within science.[39] Still it is surprising that we are quite *so* vulnerable to such influences, to the extent that the collective response has been to allow a huge increase in emissions even as the mainstream science has solidified. Indeed, one could be forgiven for thinking that the more evidence we get, the more we demand, and the more reckless our behavior becomes. In the abstract, this seems bizarre; amid the temptations of the perfect moral storm, it is sadly predictable.[40]

The second illustration involves selective attention. Since climate change involves a complex convergence of problems, it is easy to engage in *manipulative or self-deceptive* behavior by applying one's attention only to some considerations that make the situation difficult. At the level of practical politics, such strategies are all too familiar. However, selective attention strategies may also manifest themselves in the theoretical realm. For instance, consider this unpleasant thought. Perhaps the dominance of the tragedy of the commons model (and the global storm approach more generally) is not due to mere obliviousness to the tyranny of the contemporary, but is instead encouraged by its background presence. After all, the current generation may find it highly advantageous to draw attention toward various geopolitical issues that tend to problematize action, and away from issues of intergenerational ethics, which demand it. From this point of view, the standard tragedy of the commons account fits the bill nicely. It essentially *assumes away* the critical intergenerational dimension, by taking the relevant actors to be nation-states who represent the interests of their citizens in perpetuity. In addition, by focusing on countries and not generations, it suggests that failures to act count as *self-inflicted and self-destructive* harms, rather than injustices we inflict on future people.

The problem of moral corruption also reveals a broader sense in which climate change may be a perfect moral storm. Its complexity may turn out to be *perfectly convenient* for us, the current generation, and indeed for each successor generation as it comes to power. It provides each generation with the cover under which it can seem to be taking the issue seriously—by negotiating weak global accords that lack substance, for example, and then heralding them as great achievements[41]—when really it is simply exploiting its temporal position. Moreover, all of this can occur without the guilty generation actually having to acknowledge that this is what it is doing. By avoiding overtly selfish behavior, an earlier generation can take advantage of the future without the unpleasantness of admitting it—either to others, or, perhaps more importantly, to itself.

Notes

1. Barack Obama, "Inaugural Address," 2013, The White House, President Barack Obama, http://www.whitehouse.gov/the-press-office/2013/01/21/inaugural-address-president-barack-obama.
2. This chapter draws from, but goes beyond *A Perfect Moral Storm*, especially chapter 1.
3. United Nations, *United Nations Framework Convention on Climate Change*, United Nations, FCCC/INFORMAL/84 GE.05-62220 (E) 200705 (1992), 9.
4. Eric Posner, "'You Can Have Either Climate Justice or a Climate Treaty, Not Both,'" Slate (2013), November 19th.
5. Attributed by the Indian delegation. Stern's meaning is not entirely clear. In a subsequent press conference, he suggested that the United States wanted to block language that might preserve a "firewall" between developed and developing countries, but accepted that "equity" is enshrined in the UNFCCC and supported "fairness to all parties" See Johnathan Pickering, Steve Vanderheiden, and Seumas Miller "If Equity's In, We're Out." *Ethics & International Affairs* 26, no. 4 (2012): 423–443, and United States Department of State. "United Nations Climate Change Conference in Durban Special Briefing: Todd Stern" (2011, December 13th).

6. When writing their pieces, all three were at the University of Chicago Law School; after serving in the Obama administration, Sunstein is now at Harvard.

7. Other instances of the perfect storm are likely to emerge over time. Already we see degenerate cases, such as ocean acidification, long-term nuclear waste, and the dismantling of post-Depression financial regulations.

8. IPCC *Climate Change 2001: Synthesis Report* (Cambridge: Cambridge University Press, 2001), 2, emphasis added.

9. Though wary of "values" because of potential ontological implications, I use it here for simplicity.

10. Paradoxically, even if we learn that severe impacts are "in the cards" over the couple of decades, this may make matters even worse in the longer-term (e.g., if a threatened generation boosts emergency production and so emissions). See Stephen M. Gardiner, *A Perfect Moral Storm* (New York: Oxford University Press, 2011), chap. 6.

11. BBC 2009. "Copenhagen Reaction in Quotes" (December 19).

12. Stephen M. Gardiner, "Ethics and Global Climate Change," *Ethics* 114, no. 3 (2004): 555–600.

13. Sebastian Junger, *The Perfect Storm: A True Story of Men Against the Sea* (New York: W.W. Norton, 1999).

14. Garrett Hardin, "The Tragedy of the Commons," *American Association for the Advancement of Science* 162 (1968): 1243–1248; cf. Stephen M. Gardiner, "The Real Tragedy of the Commons," *Philosophy & Public Affairs* 30, no. 4 (2001): 387–416.

15. Eric A. Posner and David A. Weisbach, *Climate Change Justice* (Princeton: Princeton University Press, 2010), 42.

16. For a more sophisticated account, see Gardiner, *A Perfect Moral Storm*, chap. 4.

17. This implies that, in the real world, commons problems do not, strictly speaking, satisfy all the conditions of the prisoner's dilemma paradigm.

18. Hardin, "The Tragedy of the Commons", 1247.

19. Dale Jamieson, "Ethics, Public Policy, and Global Warming," *Science, Technology & Human Values* 17, no. 2 (1992): 139–153.

20. I discuss some psychological aspects of political inertia and the role they play independently of scientific uncertainty in Gardiner, *A Perfect Moral Storm*, chap. 6.

21. IPCC, *Climate Change 2001: Synthesis Report*.

22. Ibid.

23. IPCC, *Climate Change 2013: The Physical Science Basis* (Cambridge: Cambridge University Press, 2014), 472–473.

24. IPCC, *Climate Change 2007: The Scientific Basis* (Cambridge: Cambridge University Press, 2007), 822, http://www.ipcc.ch/publications_and_data/ar4/wg1/en/spmsspm-direct-observations.html.

25. "Tyranny of the contemporary" refers to a class of intergenerational problems that includes (for example) the pure intergenerational problem (PIP) and the generation-relative interests problem (GRIP). See below.

26. Gardiner, *A Perfect Moral Storm*, chap. 5.

27. I thank Dustin Schmidt for the acronym.

28. James Hansen and Makiko Sato, "Greenhouse Gas Growth Rates," *Proceedings of the National Academy of Sciences* 101, no. 46 (2004): 16109–16114.

29. I owe this suggestion to Henry Shue.

30. Gardiner, *A Perfect Moral Storm*, chap. 10.

31. Although it is unclear whether ultimately this should be thought of as a distinct storm, here I treat it as such to highlight some central ethical issues.

32. E.g., Paul Taylor, *Respect for Nature: A Theory of Environmental Ethics*. (Princeton: Princeton University Press, 2011).

33. Mark Sagoff, *The Economy of the Earth* (Cambridge, UK: Cambridge University Press, 1988).

34. Bill McKibben, *The End of Nature* (New York: Random House, 1989); Allen Thompson, "Virtue of Responsibility for the Global Climate," in *Ethical Adaptation to Climate Change*, eds. Allen Thompson and Jeremy Bendik-Keymer (Boston: MIT Press, 2012), 203–221.

35. Stephen M. Gardiner, "Rawls and Climate Change: Does Rawlsian Political Philosophy Pass the Global Test?," *Critical Review of International Social and Political Philosophy* 14, no. 2 (2011): 125–151.

36. Gardiner, *A Perfect Moral Storm*, chap. 8.
37. Robert J. Samuelson, "Lots of Gain And No Pain," *Newsweek*, February 2005, 41.
38. E.g., UK Royal Society and US Academy of Sciences, *Climate Change: Evidence and Causes*, 1 minute, National Academy of Sciences (National Academies Press, 2014). doi: 10.17226/18730.
39. E.M. Conway and Naomi Oreskes, *Merchants of Doubt* (New York: Bloomsbury, 2010).
40. Gardiner, *A Perfect Moral Storm*, Appendix 2.
41. Stephen M. Gardiner, "The Global Warming Tragedy and the Dangerous Illusion of the Kyoto Protocol," *Ethics & International Affairs* 18, no. 1 (2004): 23–39. doi:10.1111/j.1747-7093.2004.tb00448.x.

3

Who Are We and What Do We Want?

The American way of life is not up for negotiation. Period.[1]

I HAVE ARGUED that climate change poses a profound ethical challenge for the current generation, and especially the affluent. It is a test of who we are, including of our values, and the ability of our institutions to represent those values.

Still many in policy circles remain not merely indifferent but deeply hostile to climate ethics. This hostility is more often expressed in interdisciplinary workshops and conference hallways than in print. Moreover, it involves a number of interrelated strands of argument rarely fully developed or integrated into a consistent whole. This chapter considers the most common strands, echoes of which are found in the more elaborate position of the "Chicago lawyers."

3.1 "PURE" POLICY

The first strand seeks to isolate ethics from other disciplines. It frames climate change as a "pure policy" problem, where the relevant question is "what works," and the answers must come from science, economics, international relations, and related disciplines. Talk of ethics and justice is rejected as irrelevant, and indeed dangerous[2];

philosophy is said to lack the tools for analyzing the policy problem; attempts to accommodate justice are accused of causing past policy failures. Consequently, those with ethical concerns—philosophers, political leaders, civil society groups or members of humanity at large—are told to cede the floor to experts in the favored disciplines. Put bluntly, rather than confronting us all as the great moral and political challenge of our time, climate change becomes a matter best left to the technocrats.

In my view, the "ethics vs. policy" (or "philosophy vs. what works") framing is misleading, prejudicial, and unsustainable. It suggests two sharply differentiated camps, aligns specific disciplines with each (philosophy with one; economics, political science, international relations, natural science and engineering with the other), and cedes both the title "policy" and the label "what works" to the 'nonethics 'camp. Yet each of these moves should be resisted.

3.1.1 Marginalization by Definition

To begin with, ethics should not be ruled out of policy by definitional fiat. In my view, ethical analysis is an essential part of policy work. Understanding values and ethical parameters is a vital step in identifying the problem to be solved, providing guidance on how it may be addressed, and assessing whether proposed solutions are acceptable. Consequently, dismissing ethics would require a strong argument. Ruling it out by definition is no argument.

Worse, the "pure policy" approach disguises what is really going on. Since values are essential, the approach effectively relinquishes the ethical decisions to "technical" experts, often obscuring their value judgments behind a veil of technical language. This is not only morally and politically inappropriate, but also dangerous, especially in a perfect moral storm where the threat of corruption is high.

The "philosophy vs. what works" framing should also be rejected, since it invites two common prejudices. The first is that the ethics camp is involved in some mysterious, abstract, and otherworldly endeavor ("philosophy") that is out of touch with reality, irrelevant to action, and ultimately pointless. In context, this prejudice is ill founded. The central concerns of climate ethics are issues such as suffering, vulnerability, injustice, rights, and responsibilities. These are hardly mysterious, otherworldly, or pointless concerns. Moreover, it is difficult to see how they could be set aside in an analysis of "what works." To put the point polemically, "if climate policy ignores issues such as preventing suffering, injustice and massive human rights violations, what is it concerned with?" ("What is the point *of it*?") Given the intelligibility constraints mentioned in chapter 2, this is not an empty challenge.

The second common prejudice is that "philosophy" and "ethics" are somehow the purview of starry-eyed idealists too naïve for the real world, and so likely to bring on well-intentioned folly. However, there is nothing "starry-eyed" about the perfect moral storm analysis, or consideration of the real sufferings, vulnerabilities and responsibilities brought on by climate change. These are central aspects of the problem to be confronted (and presuppose a rather grim reality). Moreover, ambitious ideals can be found in both camps. Consider, for example, the common assumption that current national institutions can be relied upon to pursue the best interests of the next few generations, or Posner and Sunstein's background goal of maximizing global welfare. Why think such views are notably *less* politically naïve than mainstream climate ethics?

A more evenhanded contrast than "philosophy vs. what works" might be between "what works" and "what *matters*." This framing tends to valorize ethics. (How could talk about "what matters" be irrelevant, or dangerous?) It also makes it clear why ethics cannot easily be dismissed from policy. (How can you determine "what works" without considering *what matters*?)

3.1.2 Segregation

The neat assignment of disciplines to camps should also be rejected. The so-called "policy" disciplines clearly take normative positions. For instance, much work in climate economics aims at increasing overall welfare, and much in international relations at enhancing peace and security. These are value-laden concerns. Similarly many internal disputes in such disciplines concern ethics, rather than "what works." For example, arguably the most important dispute within climate economics—the one that essentially drives the vastly different policy recommendations of its practitioners—is about how to value the future. One key question in that dispute is whether climate decisions should be based on a discount rate that includes a pure time preference: the observed preference of current people to get things earlier rather than later, even when this means getting less, due to impatience (rather than, say, uncertainty or risk). Another key question is whether future people should pay more for climate change if they are likely to be richer than current people. Both questions are inherently ethical, not merely technical issues in economics.

The attempt to isolate ethics is also prejudicial. No other discipline is forced to work alone. Moreover, isolationist "climate ethics" is a straw man. No philosopher argues that climate science (for example) is irrelevant to climate ethics; and the implicit suggestion that ethicists are uninterested in "what works" is untrue. Contributions from science, economics, political science and other disciplines are vital to an ethical approach to public policy, and writers on climate ethics make much use of them.

Indeed, even the idea that practical ethics *could* operate in complete isolation from other disciplines is bizarre. Empirical information about the world is needed if ethics is to be applicable. For instance, ethical principles (such as "if empirical situation X arises, action A is the right thing to do") cannot be applied without information about whether the appropriate conditions (i.e., X) actually hold. This cannot come from ethics itself, but must originate elsewhere, and especially from other policy-related disciplines. The answer is not "ethics or policy," but "both."

One place where the prejudice against ethics becomes prominent is in the complaint that ethics lacks the "right tools" to contribute to climate policy, such as those needed to understand incentives, costs, and so on. This complaint is misleading, since climate ethics does not neglect these areas. For example, the perfect moral storm analysis is deeply concerned with incentives, and pays sustained attention to the role of game theoretic models in economics and international relations. Similarly, those advocating for particular distributive regimes, such as equal per capita emissions, typically embed them in systems of tradable permit systems ("cap-and-trade") that come from economics, and do so partly because of the incentive effects. Indeed, arguably, climate ethics typically takes incentives more seriously than standard "policy" approaches. For instance, these typically devote no attention to the tyranny of the contemporary, and instead promulgate the traditional tragedy of the commons analysis under the (in my view dangerously naïve) assumption that current institutions reliably promote the interests of future generations.

3.2 QUASI-SCIENTIFIC IMPERIALISM

If policy needs ethics, we might expect to find holes or gaps in allegedly "pure" policy arguments. This leads us to the second strand of hostility, which aims to sidestep ethics through direct appeals to science.

3.2.1 Goals

One strategy involves presenting climate goals as "derived" from science. For instance, much climate policy is based around objectives such as keeping global temperature rise below 2°C, or cumulative anthropogenic emissions below a trillion tons, and so on. Consequently, many implicitly assume that climate goals are "value free," so that the need to discuss ethics disappears from view.

Unfortunately, philosophers of science typically argue that the practice of science is not "value free" in the relevant sense.[3] The 2°C threshold for "dangerous" climate change provides a useful illustration. Procedurally, it "has emerged nearly by chance, and evolved in a somewhat contradictory fashion: policymakers have treated it as a scientific result, scientists as a political issue."[4] Substantively, the target makes implicit decisions about tradeoffs that need to be justified. For example, if two degrees would sink small island states and fry Africa, then its intelligibility as capturing the goal of avoiding *dangerous* climate change requires defense.

More generally, attempts to justify the two degrees threshold also raise ethical questions. To illustrate, consider one early rationale. Apparently, the target's first appearance came in a marginal comment by the *economist*, William Nordhaus, who thought it "reasonable to argue that the climatic effects of carbon dioxide should be kept within the normal range of long-term climatic variation."[5] This rationale is not obviously ethically defensible. For instance, suppose someone made the parallel proposal to limit the fatalities caused by a new industrial toxin to the maximum number of deaths caused by naturally occurring diseases (such as cholera). Such a benchmark would be ethically problematic for numerous reasons: it allows additions to the total number of fatalities; it assumes that we have the right to inflict deaths comparable to those occurring naturally; it does not consider our duties to reduce both natural and human-caused fatalities; etc. Nordhaus's rationale for two degrees raises similar objections. Consequently, it does not justify the threshold as a narrowly "scientific" or "pure policy" goal.[6]

3.2.2 Implementation

One strategy for preserving a two-camp approach would be to concede that while ethics informs goals, implementation is the province of other disciplines. This is a retreat from the original rhetoric: ethics is no longer irrelevant or dangerous, but plays an exalted role in setting the agenda for "what works."

Nevertheless, the "pure implementation" view remains prejudicial, since it tacitly assumes that implementation does not involve ethics. This is surely false. Just as you cannot build a good building if you do not know what it is for (e.g., housing or trash disposal), so there are ethical restrictions on how you go about building it, whatever it is for (e.g., prohibitions on stolen materials, slave labor). One plausible constraint in the climate case is that emissions reduction policies should not exacerbate injustice. Most obviously, extreme implementation methods—such as involuntary sterilization campaigns or depriving the very poor of subsistence emissions—should be off the table. A more interesting question is "what else?" In my view, policy professionals should consider such questions a vital component of their work, and therefore have significant training in recognizing ethical concerns. Without this, climate policy is in danger of violating intelligibility constraints. For example, it might advocate building small lifeboats (for the privileged) when it should be pursuing an ark (for all).

3.3 FEASIBILITY

A third strand of hostility to ethics involves protests of "infeasibility." All views accept some feasibility constraints (e.g., no one proposes magic wands). These usually involve hard physical or economic facts imposed by the world (e.g., we cannot simply *declare* that the oceans absorb much more carbon dioxide, or that the world is fifty times richer). By contrast, the economic realists' "feasibility" constraints typically rest on political claims about what *we* are incapable of, or simply will not do. These are more controversial.

First, denials of political feasibility are notoriously treacherous. In my lifetime, many things previously touted as infeasible by seasoned political commentators and conventional wisdom have occurred (e.g., the fall of the Berlin Wall, the peaceful collapse of apartheid, the election of a black President). Pessimistic predictions based on past experience are not always a good guide to the future. Sometimes we do what (morally) needs to be done regardless of history, especially when there is no other way.

Second, alleged feasibility constraints often contain disguised value judgments, and in ways that prejudice policy. For example, Weisbach suggests an "iron law" of economics correlating wealth with energy use: "we cannot be wealthy without energy" and "energy use goes up with wealth at almost exactly the same rate in every country."[7] Yet notice that insisting on the iron law as a feasibility constraint implicitly *rules out* some approaches to climate policy, such as reducing overall energy consumption, or decoupling it from well-being, or promoting objectives other than wealth (e.g., freedom, community preservation). For instance, I am sympathetic to the capabilities approach to well-being pioneered by Amartya Sen and Martha Nussbaum. Yet many central capabilities do not seem crucially dependent on high levels of material wealth or energy, but can be promoted in other ways (e.g., education, women's rights, leisure).[8] Perhaps we should investigate "iron laws of capability" instead.

Third, when applied to politics, claims of "infeasibility" tend to obscure important ethical questions. It is one thing to say that something cannot happen because it violates the laws of physics, but quite another to say that it cannot because people are not willing to do it. This is especially so when the people in question are *us*. For instance, Posner and Weisbach repeatedly emphasize that climate justice is infeasible because *Americans* will not accept its burdens. Hence, they imply that *we* are the main feasibility constraint that restricts *our own* climate policy.

While not incoherent, this claim is jarring. In my view, the underlying thought that Americans must promote unjust climate policies *because* they are *incapable* of meeting the ethical challenge is and ought to be tough *for Americans* to take. (Surely we believe we are "better than that.") Moreover, even if our detractors were right (which they are not), it would be difficult for *us* to glibly dismiss this as a "feasibility constraint," as if it were merely some inconvenience imposed by the world.[9] Arguably, such a severe moral failure would involve (as President Obama put it) betraying some of our deepest values, and so constitute a serious blow to our very conception of "who we are." Apparently, on Posner and Weisbach's account, "we have met the enemy and he is us." The technocratic language of "infeasibility" obscures this.

3.4 SELF-INTEREST

The fourth strand of hostility toward ethics maintains that we do not need ethics to tell us what to do, since self-interest already provides the answer (e.g., "stop shooting ourselves in the foot," "avoid collective suicide"). Some assume that self-interest is a nonethical value that trumps ethical ones;

others that it is a "pragmatic" feasibility constraint. The Chicago lawyers officially endorse a constraint view:

> "Any treaty must satisfy what we shall call the principle of International Paretianism: all states must believe themselves better off by their lights as the result of the climate treaty . . . in the state system, treaties are not possible unless they have the consent of all states, and states only enter treaties that serve their interests."[10]

In response, I will argue that self-interest approaches are often opaque and unstable. They frequently vacillate on core questions, obscure commitments to deeper values, and so threaten a technocratic Trojan Horse. Notably, what begins as a reasonable sounding (though highly indeterminate) "pragmatic" constraint oscillates between strictly excluding ethics, and imposing whatever values the speaker wishes. In a perfect moral storm, this is highly dangerous.

3.4.1 The Nature of the Constraint

Let us begin with how the constraint is understood. The first problem is that the quasi-economic label "Paretianism" is misleading. Traditional Pareto optimality picks out an (ethically loaded) sense of efficiency: no one can be made better off without making someone else worse off. Intuitively, this rules out a certain kind of waste: in a Pareto inefficient situation, at least one person can be made better off without making anyone worse off.

By contrast, the Chicago constraint makes the very different claim that *all* states must benefit. Call this, "International Mutual-Benefitism" (IMB). IMB has nothing to do with eliminating waste. For one thing, it is compatible with wasteful violations of Pareto optimality (e.g., a weak climate treaty might benefit everyone, but much less so than a strong climate treaty). For another, it rules out some Pareto improvements (e.g., climate policies that benefit only some [e.g., poor countries], but make others [e.g., the rich] no worse off).

The second problem is that, as stated, IMB does not specify a baseline against which the better off judgment is to be made. There are many possibilities: better off than under a climate catastrophe; better off than now; better off than if all continue to emit freely in the future; etc. Which is chosen makes a huge difference. For instance, at one extreme, "better off than catastrophe" might be easy to achieve, but of little interest to anyone (e.g., a treaty that saved one acre of the Maldives might make them *slightly* better off than losing everything, but remains unlikely to secure their consent); at the other extreme, "better off than under any alternative" might lead every nation to hold out for a deal where others do all the mitigation, and so not be feasible. Moreover, we should consider the possibility that the most relevant baseline might be ethical. For instance, countries might refuse to consent to treaties that treat them unfairly, are inconsistent with basic norms of self-respect, do not pay adequate respect to their interests (e.g., the basic human rights of their citizens, their right to escape extreme poverty), and so on.

The third problem is that the strength of IMB is radically underdetermined. Though the language of constraint suggests that self-interest limits the role of other factors, economic realists often vacillate on how sharp this limit is. First, at one extreme, self-interest might *decisively determine* what can be done (e.g., we do not need ethics to tell us to get out of the way of a moving train). Here, the "constraint" is really an absolute prohibition: ethics is irrelevant because there is no room for it.

Second, more moderate versions of IMB assert only that perceived self-interest *restricts* the role of ethics. For instance, Posner and Weisbach sometimes claim that International Paretianism (IP) "will generate a surplus—the climatic benefits minus the costs of abatement," and concede both that IP has "nothing to say" about distributing this surplus, and that distribution can be done "on the basis of *ethical* postulates."[11] The crucial question for restrictive views is how extensive the allowed zone is. If it is small, restriction is very close to determination; if large, ethics can make a substantial contribution.

Initially, Posner and Weisbach's main rationale for the surplus implies that the zone available for ethics is very large. They argue that mitigating "creates a surplus *relative to the status quo*," since "scientific evidence strongly suggests that humanity will be worse off in the future if climate mitigation efforts are not made."[12] This suggests that the relevant baseline for IMB is how well off people are *now*. Yet mainstream economic analyses suggest that the costs of climate action (at roughly 1% of global product) are considerably lower than those of future climate damages (up to 20%), and also by comparison to the benefits of future economic growth in the absence of climate change (2%-4% annually if the past is a good guide). In short, the future benefits of avoiding severe climate change are so large that not making any country worse off than it is now appears a relatively minor constraint, leaving plenty of room for ethics.

Still, elsewhere Posner and Weisbach apparently reject this optimistic scenario, insisting that ethical principles "can at best play a modest role" in treaty negotiations, and "most likely [affect] only a small portion of the surplus."[13] This suggests more restrictive (and ethically worrying) baselines. For instance, a second gloss on maintaining the "status quo" would incorporate current expectations for the future, including those based on unconstrained fossil fuel use (e.g., climate action "must make people as well off if they comply as if they cheat"[14]). Worse, a rival benchmark would have current governments seek to *maximize* their gains from a treaty, and so pursue the largest possible portion of the surplus that they can get. This may leave no room for ethics.[15] Though they officially reject maximization, Posner and Weisbach concede that they cannot rule it out, and many economic realists accept it.

Third, at the opposite extreme, a very different interpretation of IMB would regard perceived self-interest as itself *determined by ethical values*, on the grounds that ethics is essential to how we conceive of ourselves (cf. Keynes' remark that "the world is ruled by," and "practical men . . . slaves to . . . " the ideas of economists and political philosophers). This view is in serious tension with both the "ethics vs. policy" strand, and the claim that philosophy lacks the relevant tools for policy. For instance, it suggests that philosophical reflection might change a nation's perception of its interests by (for example) enabling it to see that its initial perception is wrong in light of values it already has, causing it to rethink how it understands those values, or spurring it to revise or add to those values. This is to concede a very large role for ethics and philosophy.

Of course, even here, the role of ethics remains officially constrained by the confines of a conception of "perceived self-interest" within which wider values play a role. (So we have the surprising result that "perceived self-interest" swallows ethics whole, at least from the practical point of view.) Nevertheless, this limitation may be merely verbal: if ethical values are *constitutive* of perceived self-interest, "perceived self-interest" may be sufficiently malleable to accommodate virtually any content. If so, the much vaunted feasibility constraint may be more or less empty, a rhetorical gloss that can be placed on any appropriate scheme, but one that provides no restrictions, and so no real constraint.

3.4.2 Content

A. COLONIZATION PROBLEM

In keeping with the constitutive view, when it comes to theory, economic realists usually embrace a very expansive concept of perceived self-interest. For instance, Posner and Weisbach say that altruism counts ("most people are altruistic to some extent, and it is in their interest to satisfy this altruism"[16]). Moreover, strikingly, they are willing to count *almost anything* as "altruistic," including views at the heart of climate ethics, such as "moral" commitments to "the well-being of people living in the distant future" or to "environmentalism."[17] Consequently, perceived self-interest swallows all.

This expansive account appears to involve a *colonization* of ethics; hence one might have expected the Chicago lawyers to have no theoretical objection to mainstream ethical positions. Nevertheless, Posner and Weisbach explicitly endorse the contrary position: that the expansive view rules out climate ethics. In particular, they brand their opponents "idealists" who believe that "nations should not act in their self-interest—*even in the broad sense that includes altruism*"[18]. This claim seems highly prejudicial. It threatens to paint ethics into a very (perhaps vanishingly) small box. Why suppose that climate ethics must go *beyond* arguing for altruism, especially if this *includes moral commitments to future people and to environmentalism*? Moreover, if even such moral commitments count as "self-interested," what views would not? Are there any? If so, why are only these called "ethics"?

I will return to the expansive view in a moment. However, first it is worth highlighting that in practice economic realists usually emphasize a very different, short-term, and narrowly economic conception of national self-interest. For instance, Posner and Weisbach insist that "states (and not just the United States) define their self-interest in [narrow] terms, oriented mainly toward wealth and security,"[19] that International Paretianism "probably requires that all states that participate in a climate treaty are *economically* better off,"[20] that past climate policy does "not depart from nations' perceived *short-term* self-interest,"[21] and that "most nations have only modest forms of altruism."[22] Indeed, they declare that their conception of International Paretianism *"rules out"* even the possibility of persuading the United States (and similar nations) that they have a moral obligation to bear climate burdens that reflect their wealth and responsibility. This view, Posner and Weisbach say, is based on a reading of history: "we do not expect Americans (or people in other countries) to define their national interest so capaciously because they never have in the past."[23]

The contrast between the expansive concept of self-interest and the narrow conception is stark. On the expansive, moral commitments to future generations and environmentalism are subsumed in perceived self-interest, and so might play a strong role; on the narrow, most moral argument is ruled out. Oscillation between these perspectives often makes particular economic realists hard to pin down.

The narrow view is more threatening to ethics, but also highly controversial. Personally, I find its restrictions on national self-interest unconvincing, and the empirical thesis dubious (e.g., consider the abolition of the transatlantic slave trade). Still, my main complaint is that there is an alternate explanation for any historically observed narrowness. According to my reading of the perfect moral storm, existing institutions fail to capture people's genuine intergenerational and ecological concerns, and the main ethical challenge facing our generation is to fill this institutional gap. Past failures only highlight this. Consequently, suggesting, as Posner and Weisbach do, that such failures *define* us is not merely ethically shocking, it obscures what needs to be done.

B. CONFLATION PROBLEM

Given the harshness of the narrow view, some economic realists will be tempted to retreat to the expansive view. This brings on an issue—the *Conflation Problem*—that leads us more deeply into ethical theory. To say that some action or policy X is *in an agent A's interests* is different from saying that *A is interested in* X. The latter usually means that A is positively disposed toward X (e.g., that she wants, desires, or is intrigued by X). However, to say that X is in A's interests means that X benefits A (e.g., it contributes to her well-being, or her good).

These two senses of "interest" are sometimes run together. For example, in endorsing altruism, Posner and Weisbach say:

> "Suppose that a climate treaty required the United States to lose, on net, say $20 billion, but the money would aid impoverished foreigners. To the extent that Americans are altruistic [i.e., interesting in, or positively disposed towards, altruism], the United States could consent to this treaty despite this apparent loss. *We would say that such a treaty promotes America's self-interest* [i.e., benefits Americans]."[24]

This claim generates the paradoxical result that altruism promotes the "self-interest" of those who are altruistic. Posner and Weisbach do not say why they accept this paradox. However, a popular reason[25] would be the background assumption that whenever an agent gets what she wants (i.e., is "interested in"), she is thereby benefited, because getting what one wants is the ultimate grounding of welfare. In other words, what *makes* getting X good for you is the fact that you want X (even if that X is a benefit for others). Philosophers call this "the preference-satisfaction account of well-being."

The preference-satisfaction account has many supporters. Nevertheless, we should beware of its invocation here. In general, one shouldn't move too quickly from wants to benefits. For one thing, people routinely yearn to avoid the gym, or crave that extra drink, without believing for a moment that satisfying such desires benefits them. So wants do not always correlate even with perceived self-interest. For another, our actual preferences are often badly informed, irrational, or the product of suspect processes (e.g., manipulative advertising). Hence, wants also do not automatically track real benefits.

Consequently, most philosophical preference-satisfaction theorists claim only that it benefits us to get what we *would* want if we are suitably well informed, rational, and so on.[26] Moreover, this insight fuels rival theories of well-being. These typically claim that other things, such as the quality of our mental states or character, are ultimately what matters for welfare, and that what we happen to desire is only instrumentally important, insofar as it serves these ends.

C. ROBUST SELF-INTEREST

My sympathies are with the more robust accounts of well-being. Interestingly, they suggest that economic realists may fail to take national self-interest *seriously enough*. For instance, "being in the national interest" is typically an honorific label, signaling a normative achievement that can be endorsed by ethical analysis.[27] Importantly, for something to count, it must satisfy at least some appropriate standards. Consequently, governments that genuinely pursue their peoples' long-term interests face real intelligibility constraints on how they understand those interests (e.g., some climate impacts would violate most reasonable conceptions of national self-interest, such as the United States losing Florida to sea-level rise). For instance, to my mind, we might expect ethical guidelines such as the following.

First, robust accounts of national self-interest would require governments to be much more concerned with establishing effective intergenerational institutions than with dubious short-term economic calculations. Second, over the very long term "national self-interest" may cease to have any real content independent of a sense of the *global interest* of humanity as such. (People migrate; countries break up. Over centuries and millennia, the world moves on.) Consequently, governments that adopt a rigidly nationalistic perspective may undermine their own objective, as well as their citizens' longer-term aspirations for who they are and aspire to be.

More generally, we should expect accounts of national self-interest to be ethically demanding. Even in the case of individuals, making sense of self-interest requires integrating a complex set of goals, needs, desires, and constraints. Consider, for example, asking the average high school freshman what kind of life she thinks best for her in the long term. This requires her to integrate a complex set of goals, desires, and ambitions, many of which are not yet well formed, and some of which flow from dubious assumptions and limited experience of the world. At such an age (and often much later), most of us would find the question daunting, not least because we have yet to figure out "who we are" and aspire to be.

Arguably, such problems are even more challenging when it comes to nations. The concept of national self-interest requires integrating the goals, needs, rights, responsibilities, and so on of many millions—sometimes billions—of individuals across space and time. It involves saying what matters about the lives of very many, often very different, kinds of people (and similarly diverse community projects and values) over a history that lasts hundreds— sometimes thousands—of years. On the face of it, this is not only a nontrivial task, but one likely to involve substantial moral and political philosophy. We need to ask serious questions about what matters to us, how competing values are to be reconciled, and so on.

3.4.3 Responsibility

Still, in my view, the political concept most relevant to "who we are" is not self-interest but responsibility, and in particular the ideal of generational responsibility stressed in chapter 1. There are several ways in which this ideal might be understood, but here I will sketch just one. According to a long tradition in political theory, (a) political authorities act in the name of the citizens in order to solve problems that either cannot be addressed, or else would be poorly handled, at the individual level, and (b) this is what, most fundamentally, justifies both their existence and their specific form. Political institutions and their leaders are legitimate because, and to the extent that, citizens delegate their own responsibilities and powers to them.

From this perspective, the most direct responsibility for recent climate policy failure falls on recent leaders and national institutions. If authority is delegated to them to deal with global environmental problems, they are failing to discharge their responsibilities and subject to moral criticism. Of course, on my view, such institutions were not really designed to deal with large global and intergenerational problems; hence the assignment of responsibility is to some extent unfair. Nevertheless, existing leaders and institutions have assumed the mantle of responsibility, making many fine speeches, organizing frequent meetings, promising progress, and so on. Hence many have acted as if the role did belong to them and they were capable of discharging it. They have not, for instance, bravely declared to their constituencies that the issue is beyond their competence, nor have they advocated for institutional change (e.g., a global constitutional convention). Given this, they can be held at least partly responsible.

Still, the more important issue is that, without effective delegation, responsibility falls back on the citizens, either to solve the problems themselves, or else, if this is not possible, to create new institutions to do the job. If they fail to do so, then they are subject to moral criticism. In my view, this is the challenge for the current generation, and the central threat to "who we are." One consequence is that the economic realist strategy of treating the short-termism and narrow economic focus of conventional institutions as "feasibility constraints" badly misrepresents the problem we face.

3.5 INSTITUTIONAL OPTIMISM

This issue leads to a further strand of anti-ethics thinking. Paradoxically, economic realists are often strikingly optimistic about current institutions.

3.5.1 Rival Constraints

One manifestation comes when they fail to take seriously threats posed by alternative (plausibly tougher) constraints than IMB. For instance, one rival, inspired by the tyranny of the contemporary, is *Generational Mutual-Benefitism (or Generational "Paretianism")*: each generation, when it holds the reigns of power, must believe itself better off by its own lights if it accepts a climate treaty. A second is *Elitist Mutual-Benefitism (Elitist "Paretianism")*: each national elite (e.g., the Party, the "1%") must believe itself better off by its own lights if it accepts a climate treaty. A third possibility is *Fractured Power*: a climate treaty must have at least the short-term support of a (potentially shifting) coalition that amounts to a critical mass of the politically influential. Notably, the most relevant constraints may have nothing to do with any serious conception of national self-interest. If (for example) what is really going on in some places is that various fat cats or corporations are ruthlessly competing with each other in the fleecing of countries for their own ends, then the outcome may be chaotic with respect to *any* conception of national self-interest, however short-term or economic.

Economic realists often obscure such possibilities. On the one hand, they make bold empirical assertions—such as "what is true is that states usually define their interests in terms of the well-being of their populations"—without explaining why such claims are "realistic" (rather than naïve). On the other, they sometimes define away the problem, saying that national interests "are constructed out of some aggregation of the interests of *the people who control the state*,"[28] admitting that "if a state defines its national interest in a way that takes account of the well-being *of few* of its citizens, it may enter a treaty *that complies with IP* but *has extremely bad consequences for many people*."[29]

3.5.2 Invisible Hands

Optimism sometimes reveals itself in another way.[30] When attacking ethics those attracted to self-interest usually invoke short-term economic considerations. However, in defending their own views they often appeal to much wider concerns (e.g., many invoke the children and grand-children, and so implicitly expand "self-interest" to include two or-three generations' worth of interests).[31] Yet it is far from clear that economic realists can have it both ways. Moreover, it seems highly implausible that current institutions reliably pursue the interests of the next few generations, and very bold for realists to assume that they can (feasibly) be made to do so.

One way to reconcile these two very different conceptions of national self-interest would be to posit an invisible hand, claiming that pursuit of short-term economic interest turns out to be the best strategy for promoting (say) the three-generational interest, since short-term economic growth ultimately produces benefits that trickle down the generations. Unfortunately, this argument faces major challenges in this context.

First, the invisible hand lacks the usual material with which to do its work. Climate change involves a massive externality: the biggest market failure the world has ever seen, according to the economist Nicholas Stern.[32] Given spatial, temporal, and ecological dispersion, the costs of climate change are currently not adequately registered by market prices. (For instance, future generations cannot buy up the coal to keep it in the ground.)

Second, climate change appears to be a serious collective action problem, and such problems produce the opposite of an invisible hand: situations where each agent's pursuit of her own objectives frustrates the realization of those objectives (an "invisible boot"). For instance, if severe impacts occur, there is strong reason to believe that this will undermine the accumulation of economic benefits into the future. Proponents of the invisible hand in other contexts might thus see climate change as a threat to the conditions *under which current invisible hand mechanisms work*, especially in the long-term. Thus, they should be concerned to protect those mechanisms through climate action.

In any case, intergenerational concern is probably best understood in ethical terms. One reason is that the attempt to subsume concern even for three generations under "self-interest" is dubious: the *vertical descent* ("our children and grandchildren") model seems too limited to do the necessary work. Some people do not have or care about children or grandchildren. Even for those who do, the implications may be limited. People typically have very few children and grandchildren. Consequently, concern for their immediate descendants does not automatically justify concern for the three generations *of the entire population*, whether of their nation, or of humanity as such (e.g., perhaps a large inheritance is a better bet for protecting one's own kids than climate action). More robust concern for the future needs to be more *horizontal*. This requires wider (ethical) values.

A second reason is that concern for three generations would be insufficient to the ethical challenge. Climate change has implications for thousands of years, some potentially catastrophic. For example, consider the possibility of a massive release of methane from the deep ocean, which scientists say may be more devastating than the Permian extinction, the worse extinction event in the planet's history. Such "time bombs" are difficult to address without an overt appeal to ethics.

Third, a radical expansion of perceived self-interest to three generations muddies the waters between an ethical and egoistic approach. It is egoistic in a standard sense to be concerned for one's children only insofar as they are instrumental in producing tangible benefits for oneself, such as pleasure and security in old age. However, to genuinely value them for their own sake and take their interests as partially constitutive of one's own good is something else, and less easily separated from ethics. (Aristotle, for example, is often thought to offer a formally egoistic theory, since he says that both moral virtue (including justice) and the good of one's friends are to be valued for their own sake, but also as part of one's own good. Yet Aristotle's is still a theory of ethics.)

One issue with the separation is that once one defines "one's own good" to include the good of others, valued for their own sakes, it is no longer clear what is at stake in the dispute. (As Aristotle points out, it is not the usual problem with egoism: that the egoists are selfish.) A second issue is that, if one insists that people who value their children and grandchildren for their own sakes regard this as a central part of "who they are," one should acknowledge that many also value justice and other ethical concerns in the same way. To call one "self-interest" and the other "ethics" requires a good argument. Without one, the exclusion of values like justice seems arbitrary.

3.5.3 Ethics as Facilitator

While I tend to be more pessimistic about current institutions than economic realists, I remain more optimistic about the role of ethics in facilitating change. Rather than presenting an obstacle, a more generous account of our self-understanding suggests a powerful *motivator* of climate action: since the perfect storm threatens our evaluation of ourselves, it may also inspire us to do, and be, more.

Most obviously, persistent inaction may lead future people to regard the current generation as the "scum of the Earth": a group of ruthless self-asserters, who cared nothing for the interests of others or nature.[33] Few want to be remembered that way. Less obviously, we may also be seen as merely shallow and apathetic bunglers, more to be pitied than reviled. In one respect, this is worse. Though they have bad motives, the scum are, at least, effective agents. Shallow bunglers are inept as well as destructive. This suggests further ethical motivation for climate action: to show that we are capable of (much) more.

Importantly, there is also a positive side to this ethical situation. On my account, the charge is not really that we are mere bunglers. The perfect moral storm is a profound ethical challenge, and the world bequeathed to us seems not only ill equipped to address it, but also to raise major obstacles. Consequently, success may take extreme effort, and constitute a heroic achievement. Given this, confronting climate change provides us with an opportunity not merely to avoid being seen as scum or bunglers, but also to become a truly great generation. Even valiant failure may earn us a seat at the table. Moreover, if we leave humanity much more capable of meeting global, intergenerational, and ecological challenges in general— of addressing not just climate change but other perfect moral storms as they emerge—we have a shot at becoming the greatest generation so far. If "who we are" ethically speaking forms a central part of our self-conception, this is powerful stuff.

3.6 WELFARISM

If ethics cannot be eliminated, what values should be to the fore? The next strand of economic realism—its sympathy for "welfarism"—might be read as suggesting an answer. Welfarists "seek policies that maximize people's well-being."[34] This view has great appeal for some economic realists, and many interpret national self-interest in welfarist terms. Still, these sympathies create tensions between the various strands of anti-ethics argument, threatening incoherence. For instance, at one point Posner and Weisbach officially endorse both welfarism and International Paretianism ("feasibility and welfarism are the two pillars of a successful climate treaty"[35]), and conceive of welfarism as an *ethical* position ("[the challenge is to] construct a treaty that is both ethical and feasible"[36]); hence, they implicitly disavow the pure policy and scientific imperialist strands. More strikingly, welfarism sometimes leads economic realists to *reject* International Paretianism. Elsewhere Posner and Sunstein explicitly do so, and illustrate this by saying that they support a nation preventing genocide at a modest cost to itself "even if that nation is a net loser."[37] Intriguingly, Posner and Weisbach also claim that "wealthy nations have an ethical obligation to help the poor [in poor nations]" which "goes beyond their natural charitable inclinations."[38]

3.6.1 Utilitarianism

The philosophical counterpart to the economic realist's "welfarism" is utilitarianism. Utilitarianism asserts that the right thing to do is to maximally promote well-being. This is a mainstream view in ethics. Hence, to the extent that economic realists rely on it, they are not repudiating ethics or philosophy, but rather embracing a specific position within them. (Sometimes this point is obscured by the fact that utilitarianism—the ethical doctrine—is a popular view within economics and law.)

One fundamental challenge to "welfarism" is that utilitarianism is a controversial position in ethics. There are many ways in which one might resist it, and some include positions that are quite close to utilitarianism, easily mistaken for it, and may capture much of its appeal.

The traditional utilitarian doctrine has three basic components. One is a view about value: well-being is the only thing worth pursuing for its own sake, or good in itself. (This is the view usually labeled "*welfarism*" by philosophers.) The second component ("*consequentialism*") is a view about how values should be treated: they should be maximized. The third component ("*impartiality*") is the claim that everyone's well-being matters in the same way: no one's well-being matters more than anyone else's just because it belongs to that person (e.g., the Queen's well-being is not more important just because she is a queen). The Chicago lawyers appear to accept all three claims.

Given these components, there are various ways to reject utilitarianism. First, one might accept that welfare is an important value without conceding that it is the *only* fundamental value (and so deny philosophical welfarism). Most rival ethical philosophies agree that well-being is worth pursuing for its own sake, but also claim that other things have this status too (e.g., freedom, rights, justice). Utilitarians do not have a monopoly on interest in well-being.

Second, one could agree that welfare is the only value, but deny that it should be maximized (and so deny consequentialism). Most notably, some claim that we should seek a sufficient or equal level of well-being for everyone, or give priority to those who are worst off, rather than maximize total well-being regardless of its distribution. For example, if maximizing merely increases the well-being of those already very well off (e.g., billionaires), leaving many others very badly off (e.g., the homeless), securing equal or adequate well-being for all often seems better.

A second fundamental challenge to Chicago "welfarism" is that there are a number of ways to be a utilitarian. Economic realists typically do so through (a) direct calculation of costs and benefits, and (b) within the standard economic framework of cost-benefit analysis (CBA). However, both are controversial, even among utilitarians.

Consider first direct calculation. There are many versions of utilitarianism. Direct calculation is most closely related to "act-utilitarianism," the doctrine that one should aim to maximize the net benefits of each of one's actions. In the recent history of moral philosophy, act-utilitarianism has been subject to several major objections. One is the complaint that utilitarianism neglects the individual. In focusing on total happiness, it is said, utilitarianism puts no weight on how happiness is distributed. Consequently, direct utilitarian calculation may lead to violations of what we usually think of as individual rights, and also to highly unequal distributions. For example, perhaps there are a lot of racists and they so love a good lynching that there are more overall benefits to be gained from allowing than prohibiting it; or perhaps the rich are so obsessed with their comforts and the poor so resigned to their lot that happiness is maximized by helping the rich get richer and giving nothing to the poor.

Utilitarians have a number of responses to such objections. Some simply deny that rights or equality are important values. However, many argue that a general strategy of promoting equality and respecting rights increases welfare overall, and so should be supported on utilitarian grounds. Specifically, many (on my understanding, most) philosophical utilitarians advocate pursuing the end of maximizing social welfare through intermediaries such as justice, individual rights, and democratic decision making, rather than through direct calculation. Some even reject act-utilitarianism in any form, in favor of indirect utilitarianisms, such as rule-utilitarianism: the doctrine that the right thing to do is to act in accordance with the set of social rules which would maximize happiness. Consequently, Chicago welfarism is controversial, indeed probably a minority position, even among utilitarians.[39]

3.6.2 Cost-Benefit Analysis

This problem is reinforced when we turn to the specific method of direct calculation favored by economic realists: standard economic cost-benefit analysis. Before assessing this, let us set aside a common confusion. The phrase "cost-benefit analysis" is used in different ways. First, it can refer to any kind of analysis that tries to pick out the positive and negative aspects of a given course of action or policy. Let us call this, "pros and cons analysis" ("PCA"). This usage seems deflationary. Arguably, all ethical analysis involves identifying "pros and cons" of some sort;

hence, endorsing "cost-benefit analysis" of this form merely amounts to endorsing *analysis*.

A second use of "cost-benefit analysis" refers to a method for choosing between policies based on which yields the highest net benefits. Call this "net benefit analysis" (NBA). NBA goes beyond PCA. It assumes "pros and cons" that are directly comparable in such a way that it makes sense to speak of a "net" benefit: that which is left over when the costs are subtracted from the benefits. Net benefit analysis is seriously controversial. For instance, many rights theorists would object to analyzing a situation involving a rights violation by counting the violation as a "cost" to the victim to be weighed against the "benefits" to the perpetrator. They do not, for example, analyze a lynching by weighing the cost to the victim against the benefits to the racists, and then producing an overall calculation of net benefit. Instead, they want to say that no "weighing" should be done: the rights of the victim are decisive, and whatever joy racists may get from lynching has no status.

The third form of CBA refers to a specific conception of net benefit analysis that relies on a set of distinctive techniques within contemporary economic theory: e.g., costs and benefits are to be understood in terms of "willingness to pay," as expressed in standard market prices (or proxies for them), assessed in terms of their net present value, and mediated through a standard positive discount rate. Call this, "market cost-benefit analysis" ("market CBA").

Market CBA is very far from "pros and cons" analysis. Consider a few standard criticisms. First, market CBA is very narrow, and so biased against some concerns. The focus on willingness to pay limits the benefits and costs that can be considered to those whose value can be expressed in economic terms. This skews the overall evaluation toward short-term consumption and individualistic values, rather than wider concerns, such as those bound up with communal, aesthetic, spiritual, environmental, and nonhuman values. For example, consider the outrage in the 90's when economists evaluated lives lost due to climate change in terms of foregone income, with the predictable result that lives (and especially deaths) in poorer nations counted for much less in their market CBA than those in richer nations.[40]

Second, as Mark Sagoff argues, market CBA rests on a "category mistake"[41]: reducing all values to matters of preference as measured by market prices confuses mere preferences, whose significance might perhaps be measured in terms of the intensity with which they are held, with values, whose import should be assessed in terms of the reasons that support them. For example, just as we should not evaluate mathematical claims (such as 2 + 2 = 4) by asking how strongly mathematicians feel about them (or, more specifically, how much they would be willing to pay for the rest of us to accept them), so (Sagoff says) we should not evaluate ethical claims in these ways. To do so misunderstands the point of both ethics and mathematics.

Third, despite its initial appearance of impartiality, market CBA is an essentially undemocratic decision procedure. For one thing, each person's influence is determined by how much they would be willing to pay for a given outcome. Yet this is profoundly influenced by the resources at their disposal. (For example, Bill Gates is willing to give more to almost any cause he is interested in than I am even to my favorite.) In addition, in practice CBAs are generated by "experts": people who amass market, survey, and other information, and then distill and interpret it for their audience. Since an analyst must make a large number of decisions about the scope of a market CBA, the data relevant to it, and how the results are integrated, her preferences and values play a large role in shaping the outcome.

The role of the analyst is especially important in the climate case. For one thing, most of the relevant cost-benefit information is unavailable. For example, we lack realistic market prices for events that will play out over the next several centuries, since the historical uncertainties are so large. Consequently, the role for personal judgment and values to intervene becomes very large. In light of this, John Broome, a defender of CBA in normal contexts, goes so far as to say that for climate: "Cost-benefit analysis ... would simply be *self-deception*."[42] For another thing, Broome complains that Nordhaus's CBA is based on the assumption that "everything will be much as it is now, but a bit hotter," which he finds complacent: "I think we must expect global warming to have a profound effect on history, rather than a negligible effect on national income."[43]

A fourth concern brings us to arguably the most contentious issue between mainstream climate ethics and climate economics: that of how to treat future generations. Market CBA employs a substantial positive "social discount rate" (SDR). Discounting is "a method used by economists to determine the dollar value today of costs and benefits in the future. Future monetary values are weighted by a value <1, or "discounted"." The SDR is the rate of discounting: "Typically, any benefit (or cost), B (or C), accruing in T years' time is recorded as having a 'present' value, PV of: $PV(B) = B_T/(1 + r)^T$." In public policy in general, the rates used vary, typically between 2%–10%. For climate change, traditional models employ rates of around 5% (for example, Nordhaus uses 5.5%[44] ; similarly Posner and Weisbach advocate discounting at the market rate of return).[45]

Social discount rates reduce the weight of future costs and benefits relative to current costs and benefits. Given the effects of compounding, this has profound effects in the evaluation of very long-term issues, such as climate change. For instance, at 1%, 1 benefit is equal to 2.7 in 100 years; at 5% it is 131.5; at 10%, 13,780.6. Clearly, the choice of SDR matters. For example, suppose we were trying to decide whether to pursue a project with costs of $10 million this year, and benefits of $100 million in one hundred years. With a discount rate of 5%, this project is not justified; with a rate of 1%, it is.

Discounting is the most controversial issue in climate economics. One reason is indeterminacy. Harvard economist Martin Weitzman tells us "no consensus now exists, or for that matter has ever existed, about what actual rate of interest to use," and the results of CBA on long-term projects are "notoriously hypersensitive" to the rate chosen.[46]

A second reason is dominance. For all their sophistication in other respects, economic models of climate change are largely driven by the single number they use to assess the future. For instance, critics of Nordhaus often claim that his choice of SDR effectively swamps the contribution of the other components of his model (e.g., the damage function), rendering them irrelevant.[47] Similarly, Nordhaus complains that the very different conclusions reached by his rival, Nicholas Stern, arise "because of an extreme assumption about discounting ... a social discount rate that is essentially zero."[48] Dominance is a deep problem from the point of view of intergenerational ethics. Given the theoretical storm, the idea that future people could be dealt with by a single number in a relatively simple economic equation is ethically astounding.

The third reason for the controversy is, however, the deepest: the SDR lacks a clear, overarching rationale, and the underlying issues are often ethical. In my view, discounting is primarily a practice: it is what economists *do* to costs and benefits that occur to the future in order to assign them a current value. Different rationales are offered for this practice under different circumstances. These include appeals to economic growth, pure time preference, democracy, probability, opportunity costs, excessive sacrifice, special relations, and the idea that our successors will be better off. It would be worth addressing each of these rationales independently. However, here I will merely emphasize that they *often pull in different directions*, and *all are vulnerable to serious objections* in at least some contexts, especially when assessing the long-term future.[49] Most importantly, many of the rationales, and especially the most influential, involve *ethical claims*, such as that the future should pay for climate action because it will be richer, or that not discounting at a positive rate demands an excessive sacrifice of current generations, or that respect for democracy requires accepting that the current generation is allowed to discriminate against future generations.[50] Consequently, an analyst's choice of discount rate usually rests on a variety of ethical decisions about these issues. Market CBA is far from an "ethics-free zone."

To sum up, Chicago "welfarism" is a complex and controversial ethical doctrine. Not only does it reflect philosophical utilitarianism, which many reject, but in policy settings it typically embodies a form of utilitarianism (market CBA), which many utilitarians reject. Consequently, economic realists who embrace "welfarism" cannot escape engagement with moral and political philosophy.

3.7 CONCLUSION

This chapter identified six strands of economic realist argument against climate ethics (pure policy, scientific imperialism, feasibility, self-interest, institutional optimism, welfarism). It argued that it is far from clear how these strands are to be interpreted or integrated, and many versions implicitly rely on ethics. In my view, this imposes a substantial burden of proof on economic realists to clarify their position, especially in a perfect moral storm, where the risk of moral corruption is high. In the next chapter, I turn to the final strand of hostility to ethics, the rejection of justice.

Notes

1. Statement attributed to US President George H.W. Bush, at the Rio Earth Summit, 1992.
2. Eric A. Posner and David A. Weisbach, *Climate Change Justice* (Princeton: Princeton University Press, 2010), 189.
3. For example, Helen Longino, *Science as Social Knowledge* (Princeton: Princeton University Press, 1990).
4. Carlo C. Jaeger and Julia Jaeger, "Three Views of Two Degrees," *Regional Environmental Change* 11, supp. 1 (2011): 15–26, doi:10.1007/s10113-010-0190-9.
5. Ibid.
6. These concerns do not show that 2 degrees is a bad standard all things considered, especially as a focal point. Whether it

is requires further argument that takes ethical values into account.

7. David Weisbach, "Gambling on the Climate," review of *Climate Casino: Risk, Uncertainty, and Economics for a Warming World*, by William Nordhaus, *New Rambler Review* (2015), http://newramblerreview.com/book-reviews/economics/ gambling-on-the-climate

8. Cf. Sen's famous example of the Indian state of Kerala.

9. Stephen M. Gardiner, "Geoengineering and Moral Schizophrenia," in *Climate Change Geoengineering*. eds. William Burns and Andrew Strauss (New York: Cambridge University Press), 11–38.

10. Posner and Weisbach, *Climate Change Justice*, 6.

11. Ibid., 7.

12. Eric A. Posner and David A. Weisbach, "International Paretianism: A Defense," *Chicago Journal of International Law* 13, no. 2 (2013): 355.

13. Posner and Weisbach, *Climate Change Justice*, 96. They are also dismissive for other reasons: e.g., "while distributing the surplus in favor of poor countries satisfies International Paretianism and hence *cannot be ruled out on feasibility grounds*, this approach deserves skepticism. It may lead to perverse incentives and not serve justice in a rational and effective way." Ibid.

14. Eric A. Posner and David A. Weisbach, "International Paretianism," 350.

15. Posner and Weisbach agree that a maximization strategy leaves "little room" for fairness, but nevertheless argue that a treaty would enhance welfare relative to the status quo (see "Parietanism," 355). Interestingly, they treat maximization as an objection to IP rather than an interpretation of it, though their definition of IP does not rule it out. (See also chapter 4.)

16. Posner and Weisbach, *Climate Change Justice*, 179.

17. Eric A. Posner and David A. Weisbach, "International Paretianism," 355.

18. Posner and Weisbach, *Climate Change Justice*.

19. Ibid., 357.

20. Ibid., 179.

21. Ibid., 72.
22. Ibid., 179.
23. Eric A. Posner and David A. Weisbach, "International Paretianism," 357.
24. Posner and Weisbach, *Climate Change Justice*, 179.
25. A more promising avenue might invoke a distinction between internal and external reasons (see Bernard Williams, "Internal and External Reasons," in *Moral Luck:Phiosophical Papers, 1973–1980*, by Bernard Williams [New York: Cambridge University Press, 1981], 101–113). However, I see little evidence for this in the Chicago lawyers.
26. This holds for nations, as well as individuals. Sometimes countries want what is bad for them (e.g., the British public's desire for appeasement in the run-up to WWII).
27. For instance, although the national interest may be overridden by other normative concerns, such as rights or justice, it has some ethical status—e.g., a treaty that is in every state's interest is, other things being equal, better than one that is in no one's interest.
28. Posner and Weisbach, *Climate Change Justice*, 6.
29. Eric A. Posner and David A. Weisbach, "International Paretianism: A Defense." *Chicago Journal of International Law* 13 (2013): 349 and 352, emphasis added.
30. This section draws on Stephen M. Gardiner, "The Pure Intergenerational Problem," *The Monist* 86, no. 3 2003): 481–500, and Stephen M. Gardiner, *A Perfect Moral Storm* (New York: Oxford University Press, 2011).
31. Appeals to "quasi-moral" considerations such as personal attachment are ubiquitous in climate circles among theorists of all stripes—see Michel Bourban, "Climate Change, Human Rights and the Problem of Motivation," *De Ethica* 1, no. 1 (2014): 46; Katia Vladimirova, "The Pure Intergenerational Problem and the UNESCO Decade of Education for Sustainable Development," *Ethics in Progress* 5, no. 1(2014): 71; and J. Baird Callicott, *Thinking Like a Planet* (Oxford: Oxford University Press, 2014), xx. They are also important in wider discussions of future generations—for example David Heyd, "A Value or an Obligation: Rawls on

Justice to Future Generations," in *Intergenerational Justice*, eds. Axel Gosseries and Lukas Meyer (Oxford University Press, 2009); Jane English, "Justice Between Generations," *Philosophical Studies* 31, no. 2 (1977): 91–104; Gardiner, "The Pure Intergenerational Problem," *The Monist* 86, no. 3 (2003): 481–500; Stephen M. Gardiner, "A Contract on Future Generations?" in *Intergenerational Justice*, eds. Axel Gosseries and Lukas Meyer (Oxford University Press, 2009).

32. Nicholas Stern, *The Economics of Climate Change: The Stern Review* (Cambridge: Cambridge University Press, 2007), viii.

33. Stephen M. Gardiner, "Are We the Scum of the Earth?," in *Ethical Adaptation to Climate Change*, eds. A. Thompson and J. Bendik-Keymer (Boston: MIT Press, 2012), 241–260.

34. Posner and Weisbach, *Climate Change Justice*, 171.

35. Ibid., 6.

36. Ibid., 5.

37. Eric A. Posner and Cass R Sunstein, "Pay China to Cut Emissions," *The Financial Times*, August 5, 2007, http://www.ft.com/intl/cms/s/0/e67a8166-436d-11dc-a065-0000779fd2ac.html#axzz39YR7J6pj.

38. Posner and Weisbach, *Climate Change Justice*, 174.

39. A third challenge involves rejecting impartiality. For instance, many nationalists favor the interests of their own country above (and perhaps to the exclusion of) others. This emerges in both International Paretianism and the frequent refrain in international discourse that nations will do nothing that is not in their own economic interests.

40. John Broome, *Climate Matters: Ethics in a Warming World* (New York: Norton, 2012).

41. Mark Sagoff, *The Economy of the Earth* (Cambridge, UK: Cambridge University Press, 1988).

42. John Broome, *Counting the Cost of Global Warming* (Cambridge, UK: White Horse, 1992), 19.

43. Ibid., 25.

44. William Nordhaus, *A Question of Balance: Weighing the Options on Global Warming Policies* (New Haven, CT: Yale University Press, 2009), 61.

45. Posner and Weisbach, *Climate Change Justice*, 167.
46. Martin Weitzman, "Gamma Discounting," *Academic Economic Review* 91, no. 1 (2001): 260–261.
47. Jesper Gunderman, "Discourse in the Greenhouse," in *Sceptical Questions and Sustainable Answers*, eds. Christian Ege and Jeanne Lind (Copenhagen, Denmark: Danish Ecological Council, 2002), 139–164.
48. William Nordhaus, "A Review of the 'Stern Review on the Economics of Climate Change,'" *Journal of Economic Literature* 45, no. 3 (2007): 689, http://www.jstor.org/stable/27646843.
49. Gardiner, *A Perfect Moral Storm*, chap. 8.
50. The most obvious defense of a positive SDR (r) is that it accounts for growth (g). This is often central to rhetorical defenses of discounting (e.g., Nordhaus). However, the dispute within climate economics (where, for example, Stern and Nordhaus have similar numbers for growth) is largely about other components of the SDR, such as pure time preference (δ) and consumption elasticity (η), where the issues are ethical. See Gardiner, *Perfect Storm*, 271ff.

4

Justice vs. Extortion

If equity's in, we're out.[1]

HOW CAN WE, the current generation, and especially the more affluent, rise to the ethical challenge of the perfect moral storm? How do we resist the various temptations, achieve at least minimal moral decency, and so avoid betraying the future or becoming the "scum of the Earth"? How might we go beyond this to be a genuinely great—the greenest—generation?

Our task would be easier if we had robust theories to guide us—especially concerning intergenerational ethics, global justice and community, and humanity's relationship to nature. Alas, the lack of such theories is itself a major part of the challenge (and indulging inadequate theories one of the major temptations). What we need then is a sense of what to do in their absence. Fortunately, there are ways forward. For one thing, there are clear violations: "it may often be clear that a suggested answer is mistaken even if an alternative doctrine is not ready to hand."[2] For another, we can discern a general direction for policy through devices such as overlapping consensus and reasonable constraints. This chapter explores such strategies in light of economic realist objections. I begin with two strategies that seem deeply worrying in the context of the perfect storm, since they appear to encourage climate extortion, a clear ethical violation.

4.1 GLOBAL EXTORTION

4.1.1 "Polluted Pay"

The final strand of economic realist argument is its hostility to justice. Such hostility comes at a price. It puts economic realists at risk of endorsing strikingly counterintuitive conclusions about international climate policy. Consider, for instance, the Chicago lawyers:

> Like it or not, the *only way* for other nations to ensure Chinese cooperation is through a special inducement, such as cash or extra emissions rights. Here is the harder question: *should the United States also be paid for its participation? No one is suggesting such an approach and this should be puzzling.*[3]

> [An optimal climate treaty] could well *require side payments to rich countries* like the United States and rising countries like China, and indeed possibly *from very poor countries* which are extremely vulnerable to climate change—such as Bangladesh.[4]

> Suppose, as seems clear, that India and Africa would pay little and gain a great deal from an agreement, whereas the United States would pay somewhat more and gain somewhat less ... the standard resolution of the problem is clear: the world should enter into the optimal agreement, and the United States should be given side-payments in return for its participation.[5]

In short, the Chicago suggestion is that the most vulnerable countries (primarily, poor and low-emitting nations such as Bangladesh) should *pay off* the large emitters (e.g., the United States and China) to stop emitting so heavily.

This "polluted pay" (and also "polluters get paid") principle will strike many as outrageous. It also contrasts dramatically with the more familiar "polluter pays" principle in international law. Nevertheless, it cannot be dismissed as simply a peripheral feature of the Chicago view. Instead, "polluted pay" appears central to their basic logic and proposed solution, as the *core mechanism* through which they confront "International Paretianism" (IP): their demand that all nations benefit from climate policy.

Consider that, according to their favored cost-benefit analyses, poor countries have *the most to lose* from negative climate impacts, since they are much more vulnerable (largely because they are poor) than the richer, big emitters (since, being rich, they are better able to cope). For instance, Posner and Sunstein emphasize that standard CBA shows that the United States, China, and Russia have relatively little to lose, "the poorest nations will be the biggest losers by far," and India and sub-Saharan Africa specifically will be "massive losers."[6] Consequently, the critical thought is that poor countries should be willing to "compensate" the big emitters for making cuts that are larger than the big emitters themselves would choose in their own self-interest. Such a transaction makes sense because it makes both rich and poor nations better off than they would be otherwise, if global emissions continued to rise unchecked. Crucially, without polluted pay, the allegedly "pragmatic" policy approach will not work to check dangerous climate change. If the big emitters are driven solely by narrow self-interest, there will be a shortfall in their mitigation efforts from the point of view of the more vulnerable. On the Chicago approach, the more vulnerable must find some way to make meeting this shortfall in the interests of the big polluters; hence some form of "side-payment" is absolutely required.

4.1.2 Presumptions Against "Polluted Pay"

"Polluted pay" is a deeply problematic approach to climate policy. First, it may render ambitious climate action *infeasible* even if International Paretianism holds. For one thing, perhaps the very vulnerable do not have anything that powerful high emitters want enough to stop overemitting—given their poverty, they may have less to offer the rich than the rich could gain from continuing to overemit. For another, the most vulnerable are often relatively weak; hence, perhaps rich nations can simply take whatever interests them by other means.

Second, there are deep ethical objections. The demand that poor nations (e.g., Haiti, Sudan, Afghanistan) make "side payments" to "compensate" richer countries (e.g., the United States, China) misses central features of the climate problem, such as that the high emitters are predominantly causing it, that they have no right to do so, and that the poor countries are largely victims. Under such circumstances, it seems *morally indecent* to demand that the poor "compensate" high emitters for stopping. Moreover, to say that they should accept this *because* they are much more vulnerable—largely *because* they are so poor—seems both hopelessly wrong and to compound the initial injustice.

To make this worry more vivid, notice that "polluted pay" appears to endorse *extortion*. Extortion involves obtaining something through the inducement of a wrongful use of force, threat, intimidation, or the undue or illegal exercise of power. In the paradigm case of an extortion racket, a criminal gang threatens its victims with violence unless it is paid "protection money," as "compensation" for holding back from violence. Such rackets are morally indecent because extortionists have no right to impose, and their victims have strong rights against, these threats. In such a setting, "compensation" for the gang is an ugly euphemism. The extortionist's demand is ungrounded, and therefore illegitimate. There is no "loss" that ought to be made up. Full deployment of one's threat advantage over others is not an appropriate baseline against which to claim compensation.

This is not to deny that extortion is sometimes "what works." From the perspective of the extortionist, concerned only with narrow self-interest, it clearly has benefits. Many organizations (e.g., the Mafia, totalitarian regimes) have practiced extortion with a fair degree of success. Similarly, from the perspective of the victims, acceding to extortion is sometimes the best option available. (It costs you the ransom, but at least you get your daughter back.)

Still, this stark, "pragmatic" approach to extortion seems to miss much of what is going on, especially from the ethical point of view. We simply do not view extortion as merely "a mutually beneficial transaction that enhances the welfare of both parties, under the constraint of each agent's pursuit of its own self-interest." Instead, we condemn it.

One reason is the failure of basic respect. Extortionists exploit advantages they have over others in ways that are not merely illegitimate, but deeply disrespectful. Similarly, victims are made to accede, and so to at least a minimal extent "consent," to a practice that denies them basic respect. Although they typically gain some welfare benefit, this process is inherently degrading and humiliating.

Presumably this is why some people refuse to cooperate with extortionists, even when this results in losses for them, and why we often admire them when they do so (e.g., in Mafia-style cases). Sometimes we think it better to be worse off in welfare terms than to allow our basic status as equal moral agents to be compromised. Self-respect is a factor here; but so is social recognition of our status as beings worthy of respect.

A second set of reasons to condemn extortion involves its wider social costs. Conventional wisdom suggests that paying off an extortionist encourages further extortion, in a downward spiral, so that often the best strategy is to refuse at the outset. More generally, a system that endorsed extortion would threaten many beneficial social practices (e.g., through feeding resentment and corroding the foundations of cooperative social ventures, such as trust). This is one reason why corrupt societies struggle economically; their institutions are undermined by persistent illegitimate threats to citizens and business.

Such concerns are highly relevant to climate change. Vulnerable nations may fear that high emitters will renege on their commitments once "side payments" are made, and demand more. Moreover, even if paying off the powerful works for climate, the vulnerable may worry that this only encourages them to seek further ways to demand "money with menaces" by creating new threats. Such fears foster an international climate of distrust and resentment that is destructive for wider cooperation, overall welfare, and much else.

A third reason to condemn extortion is that for many people, and especially certain kinds of societies, to be cast as an extortionist conflicts with their most central values, and especially with their conception of "who they are, morally speaking." For such agents, it is hard to see an extortion racket in a neutral way, as merely an effective way of pursuing one's own interests. Even if the material benefits of being in the Mafia, for example, are substantial, many people cannot find it within themselves to endorse such a life—to want to *be* "*that guy* (or gal)." Similarly, even if practicing extortion would benefit America (for example) materially, and is what other countries would do, Americans might reject it as being incompatible with who they are, or aspire to be: a deep betrayal of our ideals that makes the material benefits seem like "dirty money," and so not real benefits at all. Although we don't deny the power of "you can keep your money or your daughter, not both," *Americans* (like most other peoples) do not aspire to make such threats.

4.1.3 Outsourcing Justice

Presumably, the Chicago lawyers do not intend to condone extortion. Still it is unclear how they propose to avoid it. This is partly because in their writings the threat is obscured by three factors. First, they do not straightforwardly declare *who* should pay off the big emitters in the kind of ambitious treaty they claim to favor. This results in a (strategically helpful?) ambiguity: while the basic logic of International Paretianism implies the most vulnerable should pay (as the quotations attest), the background appeal to global welfare leaves room to argue that *other rich nations* ought to "compensate" the United States and China for altruistic or ethical reasons, in order to promote a more robust climate policy.

The latter approach is, of course, more morally appealing than extorting the most vulnerable. Yet, since it involves a major concession to ethics, it is not clear that Posner and Weisbach in particular can endorse it. One reason is that they explicitly claim that the zone available for ethics likely covers "only a small portion of the surplus," implying that International Paretianism is a tight constraint and the realm for "polluted pay" large. A second reason is that, if welfarism must come to the rescue anyway, Posner and Weisbach would have to explain why it cannot be invoked more directly—to imply that the big emitters ought to take on the burden of the shortfall themselves, and give up any thoughts of being "compensated" by anyone. A third reason is that economic realists would have to explain why they believe that the idea of the big emitters (e.g., China, the United States) being paid off by their ethically motivated geopolitical rivals (e.g., Russia? Japan? Europe?) is *more politically feasible* than a triumph of ethical motivation within the big emitters themselves.

The second obscuring factor is that the policy focus of the Chicago lawyers is heavily on rapidly emerging economic powerhouses (e.g., China, India, and Brazil). By contrast, the most vulnerable countries (e.g., Bangladesh, Haiti, Sudan, Zimbabwe, Nepal, Sierra Leone, Ethiopia, Afghanistan) are barely mentioned, so that their plight under "polluted pay" remains largely hidden. Unfortunately, in a perfect moral storm neglect of their (very different) perspectives is sadly predictable (see also section 3.2a).

The third, and perhaps most important obscuring factor, is that the Chicago lawyers attempt to blunt distributive criticisms of their climate policy by endorsing a "two-track" approach: climate should be dealt with by one treaty, and distributive matters by another, "foreign aid," treaty. This approach, they suggest, will lead to the most efficient pursuit of global welfare.

Unfortunately, the two-track approach has a credibility problem. First, the basic idea seems far-fetched. Why suppose that the most powerful nations, having resisted climate justice (and instead demanded an unjust climate deal) will then turn around and endorse a robust global welfare treaty? Why is this suddenly politically "feasible"?

Second, the approach may actually make benefiting the poor *more difficult* by changing the relevant baseline. Consider a simplistic illustration. Suppose the United States starts with one hundred resources and Haiti ten. If fifteen is the minimum threshold for decent social well-being, the US should (morally) give Haiti five in aid (5% of its resources). However, if Haiti first gives the US five to act on climate, the United States has 105 and Haiti five, and the United States must now supply ten in aid (9.1% of its new resources). Presumably, since the United States must sacrifice more (both relatively and in absolute terms) from its new position, the (alleged) self-interest constraint bites harder, making decency harder to achieve.

Third, the two-track approach encourages worries of a "bait and switch." Perhaps it is easy to persuade the affluent that dealing with poverty is more important than climate when they have no intention of doing much about either. If so, the "foregone" opportunity of "more efficient" poverty reduction is spurious, except to facilitate moral corruption. Consider, for example, the worrying equating of the "poverty track" with foreign aid. At 0.2% of gross national income, current US aid is far below the level recommended by the United Nations (0.7%), and less than many other countries (e.g., Sweden and Norway at 1%). It is also frequently targeted at strategic objectives rather than those most in need (e.g., in 2012, the biggest recipients were Afghanistan and Israel). Given this, outsourcing climate justice to a hypothetical foreign aid agreement is likely to extinguish it as a serious policy objective. Consequently, even Peter Singer, arguably the world's most serious advocate for assisting the poor, suggests that climate action may be the best available strategy, even if it is second-best overall.[7]

In conclusion, the presumptions against the Chicago approach are strong. In principle, it supports an extortionate approach to climate policy; in practice, the apparent remedy—outsourcing ethics to a global welfare treaty—seems dangerously naïve.

4.2 INTERGENERATIONAL EXTORTION

Threats of extortion also arise in the temporal dimension. Several writers suggest engaging in mitigation now, but passing the costs on to future generations. For instance, John Broome argues that conventional economic analysis clearly shows that efficient climate action is possible "without sacrifice" by the current generation if governments finance climate policy through increased borrowing from the future.[8] The idea is that emissions cuts benefit the future and the current generation receives "compensation" for making them, satisfying what we earlier called "intergenerational mutual benefitism" (or "Intergenerational Paretianism").

Broome advises economists to develop new mechanisms to facilitate the cost transfer.

Broome acknowledges that efficiency without sacrifice is "seriously unjust," saying it amounts to "bribery," and the "no sacrifice" baseline is wrong. Nevertheless, he suggests pursuing it has "the moral purpose" of pushing forward action. In the longer term, endorsing efficiency without sacrifice might help move us towards efficiency with sacrifice. Although endorsing "efficiency with sacrifice" now would be better, it would be a strategic mistake to make "the best the enemy of the good."[9]

I am not so sure. First, talk of "bribery" is misleading. Since future people are not yet in a position to offer bribes, the idea behind "efficiency without sacrifice" is really that current people borrow against the future *in the name of* future people. The morally suspect category is thus not bribery as such, but something more like theft or extortion. It is theft if future generations would not endorse the deal. It is extortion if they would do so only under duress (e.g., due to the severe climate threat we illegitimately impose on them). Given such possibilities, the "bribery initiated by future generations" framing arguably sugarcoats what is morally at stake.

Second, a "borrow from the future" strategy poses fresh threats to future generations. Notably, Broome's call for innovative policy instruments to expand the potential for "borrowing" opens up new avenues for intergenerational buck-passing through the (further) running up of intergenerational debt. In a perfect storm, this seems like attempting to "bribe" a mobster by offering him new weapons. ("What could possibly go wrong?") For instance, what is to stop the current generation from taking the money but still doing very little to address climate change? What is to stop it from doing so, and then coming back for more? (Indeed coming back for more when things worsen seems likely to be a profitable strategy: the more the threat increases, the more future generations should be "willing" to pay to prevent it. Given the stakes, the fact that past bribes have not worked may not be sufficient reason not to accede again.) In short, letting the economists loose on intergenerational debt may merely increase the potential for intergenerational extortion. In a perfect moral storm, this would be sadly predictable.

Such worries are not assuaged by Broome's background rationale for "efficiency without sacrifice"—that since climate damages are a negative externality, standard economic theory implies that *in principle* it should be possible to make some better off while making none worse off (a Pareto improvement). For there are many possible Pareto improvements to choose from in climate policy, and some are extortionate. For instance, the current generation may agree to climate mitigation only on the condition that it claws back *almost all the benefits* through extra borrowing, so that future generations receive very little of the net benefits available (e.g., the current generation passes on crippling intergenerational debt, but limits future warming only to 4°C). In the perfect storm, "*efficiency with extortion*" seems a live threat.

In conclusion, a strategy of intergenerational cost transfer does not guarantee even a minimally decent climate policy, and may make matters (much) worse. It also creates new threats, including of further extortion. Again there are clear cases of ethical violations, and high risks of moral corruption.[10] We neglect justice at our (actually, usually *their*) peril.

4.3 JUSTICE NOW

A genuinely green generation would practice neither global nor intergenerational extortion. Moreover, ethical concepts such as respect, fairness, and rights—concepts strongly related to the more general notion of justice—seem central to avoiding them. The relevance of justice is unsurprising. Though only one part of ethics, it is often exalted as "the first virtue of social institutions," where this implies that unjust institutions ought not to be tolerated except to avoid greater injustice.[11] Furthermore, even those who doubt that it deserves quite such preeminence (e.g., indirect utilitarians), usually agree that justice remains a central concern.

4.3.1 The Burden Claim

Given this, can we move beyond clear violations, to claims likely to be the subject of an overlapping consensus? As a rough heuristic, let us assume that the basic questions of climate policy are: where to set a global ceiling on emissions, understood as a long-term trajectory consisting of a set of constraints at particular times ("the trajectory question"); how to distribute the emissions allowed under that ceiling at a particular time ("the allocation question"); and what to do about unavoided impacts (the "impacts question"). Of these, the allocation question has received by far the most attention so far from writers on justice. Although many proposals have been made, there seems to be a broad ethical consensus that richer, more developed countries should shoulder most of the burden of action, at least initially ("the burden claim"). Moreover, there is the sense that the agreements between rival views here are more politically important than their disagreements, because (it is assumed) almost any ethically-guided policy will take us in the same general direction ("the convergence claim").

The basic grounds for these claims are readily apparent. Following Peter Singer, we might say that, at least at first glance, mainstream views of fairness all appear to support the consensus that developed countries should take the lead.[12] First, historical theories do so because the more developed countries are responsible for the majority of cumulative emissions. For instance, from 1890–2007, the United States accounted for 28% of emissions; the E.U., 23%; Russia, 11%; China, 9%; and India, 3%.[13]

Second, theories sympathetic to basic moral equality between persons support the burden claim because the developed countries produce many more emissions per capita than developing countries. For example, in 2010 average global emissions were 4.9 metric tons of carbon dioxide per capita. However, the United States average stood at 17.6, the U.K. at 7.6, China at 6.2, India at 1.7, Bangladesh at 0.4, and Haiti at 0.2.[14,15]

Third, theories that prioritize the interests of the least well off endorse the consensus because developing countries are generally much poorer than developed countries. In 2013, average per capita income in the United States was $53,101, in the United Kingdom $37,307, in China $9,844, in India $4,077, Bangladesh $2,080, and Haiti $1,710.[16] Moreover, averages conceal some of the worst problems. For instance, in 2010, "21% of people in the developing world lived at or below $1.25 a day [the extreme poverty level]" unable to meet their basic needs. This translates to 1.22 billion people, more than 15% of the world population.[17]

Finally, utilitarian theories appear to support the consensus because taking the previous considerations seriously seems likely to promote happiness. For example, Singer[18] argues that "polluter pays" (as opposed to "polluted pay") principles help to internalize incentives, that principles of equality help to reduce conflict, and that resources produce more well-being in the hands of those with little.[19]

4.3.2 Objections

Economic realists raise several objections to allocative justice.

A. ACTUAL REDUCTIONS

To begin with, they sometimes interpret the burden claim as applying to actual reductions in emissions, and object that this effectively rules out justice because the physical constraints on action have increased dramatically over the last decade, so that some developing nations will have to constrain their emissions quickly and significantly if serious climate change is to be averted.

In response, let me make three points. First, the currency of burdens need not be emissions. When emissions cuts must occur in poorer countries, richer countries can still shoulder most of the initial burden (e.g., through technology transfer, financing).

Second, the idea that rapidly developing countries such as China and India cannot grow their emissions indefinitely has always been mainstream in both climate policy and climate ethics. Although the pace of change since 1990, especially in China, caught most policy analysts (and the ethicists who relied on them) off-guard, the thought that some transition would be needed has long been a background assumption in climate circles. In particular, although the first commitment period under the UNFCCC (the Kyoto Protocol, which ran until 2012) did not demand cuts from either the now-emerging economies or from those yet to emerge, it was always assumed that many would need to restrict emissions in subsequent commitment periods. (For instance, note that China now exceeds early equal-per-capita limits, such as Singer's 1 metric tonne of carbon [3.67 tonnes of carbon dioxide].)[20]

Third, though tightening physical constraints put pressure on the old categories of "developed" and "developing," they do not (yet) undermine the basis of the burden claim. The rise of countries like China, India, and Brazil (in economic, geopolitical, and polluting terms) makes a difference to what needs to be done, and by whom, by creating an interesting intermediate category, rather than exploding the core ideas underlying the consensus. Indeed, even now (though perhaps not for long) what to say from the point of view of justice about many other nations remains essentially the same. For example, the United States, Canada, and Australia remain in the group required to take greater burdens, whereas Bangladesh, Sudan, and Afghanistan remain in the group needing to do less. Bangladesh, for example, emitted at 0.40 tonnes of CO_2 per capita in 2010. This rate is *already lower* than that suggested by the 50%-80% cut in current average global emissions many recommend by 2050 (i.e., from 4.9 to 2.4–1.0). Many other very poor nations have even lower emissions, such as Sierra Leone (0.3 per capita), Haiti (0.2), Nepal (0.1) and Ethiopia (0.1).[21] In short, many of the most vulnerable countries are *already complying* with very stringent climate goals.

A more pressing worry for the ethical consensus concerns the future stability of the convergence claim. Presumably, moving forward, the differences between distinct approaches to allocative justice, and attempts to integrate them, will have significant practical implications for specific actors, and so complicate efforts to make progress. Consequently, as well as an ethical consensus on the initial tendency of allocation policy, we will ultimately need a further convergence on what policies count as "fair enough" to particular stakeholders to provide a stable basis for ongoing action. This suggests some kind of pragmatic resolution of the allocation issue, but one that *integrates justice concerns, rather than rejecting them.*

B. DISAGREEMENT

The second common objection to allocative justice suggests this cannot be done: "there are just too many conflicting views about who deserves aid and who should pay for it."[22] Still this argument is too quick. First, the perfect storm analysis predicts disagreement purely on strategic grounds, since it is highly convenient for the current generation.

Second, while there would probably be some disagreement even among the ethically well motivated (cf. the theoretical storm), it should not be insurmountable. One reason is the early ethical consensus; another is the pressure to act for the sake of future generations and the rest of nature, and the special responsibility that gives the current generation. Indeed in view of what is a stake, I strongly suspect that the ethically well motivated could construct an acceptable form of "rough justice" that would serve the purpose. For instance, Thomas Schelling argues that our one experience with an allocation of this magnitude is the post–WWII Marshall Plan. There, he says, "there was never a formula . . . there were not even criteria; there were 'considerations' . . . every country made its claim for aid on whatever grounds it chose," and the process was governed by a system of "multilateral reciprocal scrutiny," where the recipient nations cross-examined each other's claims until they came to a consensus on how to divide the money allocated, or faced arbitration from a two-person committee.[23] Though not perfect, such a procedure did prove workable. Given the urgency of the climate problem and the large theoretical issues involved in devising a perfectly just global system, this is encouraging. Perhaps a roughly fair system can see us through in the short- to medium-term, even if over time pressure for something better can be expected to grow.

Third, my biggest reason for rejecting the disagreement objection is that it applies even more forcefully to economic realism, which (as I argued earlier) seems particularly vulnerable to the theoretical storm. One sign of this is that attempts to apply market CBA to climate change produce severe policy disagreement, of orders of magnitude (cf. carbon taxes of $7–$350 per ton), feeding "cost-benefit paralysis." Moreover, there is no escaping potential moral disagreement, since many of the underlying issues within CBA are ultimately ethical.[24]

C. POLITICALLY UNMANAGEABLE TRANSFERS

The third argument against allocative justice is that it would involve a "redistribution" or "transfer of wealth" of billions of dollars to the less developed nations. Here there are significant framing issues.

First, from the point of view of justice, the language of "*re*distribution" and "transfers" is prejudicial. It implies that the baseline against which the distributive implications of a climate treaty are to be assessed is the rich countries' current income and wealth, based in a carbon economy, together with their expectations that this pattern will continue in the same way moving forward. Yet the rich are not *entitled* to this future trajectory. Thinking this way embeds a bad baseline, and ignores the role of justice in establishing a new one. Consider, for instance, that we would not describe the abolition of the transatlantic slave trade as a mere "redistribution," nor are we paralyzed by the thought that it involved large "transfers" of wealth from the slave traders to the slaves. Similarly, no one would object to breaking up of an extortion racket because it involves a "transfer of wealth" away from the Mafia.

Second, it makes a great difference whether one frames costs in absolute or relative terms. In absolute terms, billions of dollars sounds like a lot; but against the backdrop of the global economy, it may not seem so salient, especially given the alternative. As Schelling once noted about the (much larger) costs of combating climate change:

> The costs in reduced productivity are estimated at two percent of GNP forever. Two percent of GNP seems politically unmanageable in many countries. Still, if one plots the curve of US per capita GNP over the coming century with and without the two percent permanent loss, the difference is about the thickness of a line drawn with a number two pencil, and the doubled per capita income that would have been achieved

by 2060 is reached in 2062. If someone could wave a wand and phase in, over a few years, a climate-mitigation program that depressed our GNP by two percent in perpetuity, no one would notice the difference.[25]

More pragmatically, there may be ways to make the difference seem even less salient. For instance, many changes that powerful countries could make in the name of climate justice would likely result in *net gains for all parties* over the longer term. For example, they could amend trade rules that discriminate against poorer countries, or roll back heavy domestic subsidies to agriculture that effectively deny those countries access to major markets. Since such justice-led "redistribution" proposals would be in keeping with the spirit of "International Paretianism," I suggest they are a better focus for economic realist views than "polluted pay (and polluters get paid)."

Finally, even if there must be some explicit transfers, we should not forget that the world does not stand still: life moves on and new distributions emerge. Carbon taxes or tradeable permits, for example, would provide strong incentives for those initially spending a lot on carbon to find better ways to run their affairs. Some may welcome this, for ethical and other reasons.

D. ISOLATIONISM

The fourth objection to allocative justice, and the one most emphasized by the Chicago lawyers, is "climate change blinders": wanting "each individual policy to achieve distributive goals rather than achieve redistribution through the overall set of policies."[26] This objection is more difficult to assess, partly because there are several versions.

One version of the blinders objection rests on a narrow interpretation of "distributive justice," as concerned solely with improving the global allocation of resources *as such* independent of climate change, or reflecting the idea that "rich nations have a special obligation to deal with climate change . . . simply because they are rich."[27]

The first problem with the narrow interpretation is that though some (especially "welfarist") accounts of climate justice take this form, many do not. Most notably, many justice theorists believe that the true rationale for, and objective of, distributive justice is respect for persons. Economic realists sometimes marginalize this point of view, saying that in the end climate policy is merely about money. However, this is deeply contentious. Imagine, for example, a paradigm extortionist (e.g., a Mafia kidnapper) saying this (or the more familiar "it's not personal, just business"). Chances are the victims do not agree.

Second, this dispute makes a difference to how we think about the "goal" of justice. Economic realists often presuppose that a goal is something to be "promoted," that one should strive to maximize its achievement in the world. However, many think of justice as a value to be honored or respected, rather than maximized. For instance, respect theorists typically reject the claim that pursuit of justice licenses violating the rights of a minority to promote the rights of a majority. To invoke a classic example, they insist that a small town sheriff in the Old South would not be justified in executing an innocent black man just to prevent the riot that would likely follow from refusing. Respect for justice, they say, sometimes requires failing to promote "the goal of justice" understood in a maximizing way.

This general issue reappears in the second version of the blinders complaint. Economic realists sometimes claim that distributive justice *demands* that rich countries choose the option or set of options that provides "the most bang for the buck," or best promotes well-being. Yet these claims strike me as false, and exhibiting what we may call *efficiency blinders*, especially in relation to justice.

Consider some examples. Suppose you are charged with fairly distributing pieces of cake at a child's birthday party. Would justice *demand* that you employ some maximizing criteria (e.g., give Grandpa an enormous slice, since he likes cake most)? To me it does not even *suggest* this. Or suppose you are the executor of your aunt's estate. Would justice *demand* that the neediest relation get the lion's share? (Would the fact that she loved all her nieces equally be *irrelevant*?) Again, I think not. Consequently, those concerned about justice should not quickly capitulate to the economic realists' sole and overriding concern for maximizing welfare.

A third version of the blinders complaint is more challenging. It insists that climate justice cannot be isolated from wider concerns in global ethics. I have two reactions. First, I see the relative isolation of climate justice as a feature of the early ethics of the transition rather than ideal theory, and believe that it has historical routes. Within the original policy context of climate ethics—negotiations for the Rio Earth Summit, UNFCCC, Kyoto—one allocative proposal seemed likely to dominate: grandfathering national emissions. In response, several commentators raised ethical objections, and proposed alternatives (e.g., equal per capita with trading; protecting subsistence emissions; greenhouse development rights). However, on my understanding, most intended these as *improvements* on grandfathering within a particular constrained context, and so as claims in the ethics of the transition.

Still, secondly, in my view it remains an open question whether relative isolation is a promising strategy even within the ethics of the transition. On the one hand, a climate policy that does not require too radical a restructuring of the current global system may have advantages in the short term. On the other hand, more radical theoretical and institutional approaches may be necessary. Most political theorists working today believe that the current global order is profoundly unjust, and in ways that undercut morally sensitive climate action. Moreover, my own view suggests that a serious institutional gap exists, especially when it comes to future generations, and that current institutions are in some ways hostile to intergenerational concern. Shortly, I will pursue such matters; first, however, let me say something about the more familiar area of corrective justice.

4.4 CORRECTIVE JUSTICE

Unavoided impacts raise many questions of justice. Prominent among these is the controversy surrounding past emissions. This dispute cuts across the so-called "ethics" and "policy" camps. Notably, the view that corrective justice should be applied to climate change is a minority position in climate ethics: most maintain that responsibility for past emissions is limited to post-1990 emissions, and for similar reasons to the Chicago lawyers. Still, I do not mind defending "corrective justice," since this is one place where I disagree with the majority.

4.4.1 Three Presumptive Arguments

I will focus on three kinds of ethical argument. The first appeals to *causal* principles of moral responsibility, invoking commonsense ideals such as "you broke it, you fix it" and "clean up your own mess." The core idea is that, other things being equal, those who are causally responsible for creating a problem have an obligation to rectify it, and also assume additional liabilities, such as for compensation, if the problem imposes costs or harms on others. Such principles are familiar in general, and prominent in international environmental law where (among other things) they form the basis of various "polluter pays" principles.

Another kind of argument appeals to *fair access.* Suppose that environmental services, such as the atmosphere's capacity to absorb greenhouse gases without adverse effects, are limited resources that should be shared or held in common. Given this, and other things being equal, if some agents use up the resource, and in doing so deny others access to it, then compensation is owed because the latecomers have been deprived of their fair share. This "appropriator pays" approach has similarities to polluter pays. However, there are also potential differences (e.g., "polluter pays" often suggests that at the heart of the matter is a further and independently identifiable cost or harm—a "pollution").

A third kind of argument for the relevance of past emissions rests on the net benefits of establishing a *general practice* of liability. For instance, such a practice often makes potential polluters more careful about their activities, and so reduces unnecessary costs. This may be so even if occasionally some polluters are held responsible in ways that violate other ethical desiderata (such as perfect fairness).

Sometimes the importance of the benefits outweighs the ethical violations (e.g., if the violations are much less serious, or can be dealt with in another way).

These three kinds of argument each provide a strong *prima facie* reason for regarding past emissions as *relevant* to future climate policy. Although none establishes that past emissions are the only or decisive consideration, each constitutes a *significant presumption* in favor taking them seriously. Moreover, they are in principle compatible and potentially mutually reinforcing, so that one need not necessarily pick between them. Again, there is potential for an overlapping (perhaps rough) consensus. In my view, the main objections to taking past emissions seriously begin by acknowledging this presumption, but then aim to rebut it.

4.4.2 Excusable Ignorance

The first objection claims that past polluters were excusably ignorant. They neither intended nor foresaw the effects of their behavior, and so ought not to be blamed, and cannot be held liable. As Todd Stern put it at the Copenhagen meeting: "I actually completely reject the notion of a debt or reparations or anything of the like. For most of the two hundred years since the Industrial Revolution, people were blissfully ignorant of the fact that emissions caused a greenhouse effect. It's a relatively recent phenomenon."[28]

One problem is that such claims are factually dubious. Most starkly, in 1965 the Johnson administration issued a report to Congress that concluded:

> Through his worldwide industrial civilization, Man is unwittingly conducting a vast geophysical experiment. Within a few

generations he is burning the fossil fuels that slowly accumu-
lated in the earth over the past 500 million years. The CO_2 produced by this combustion is being injected into the atmosphere;
about half of it remains there. The estimated recoverable
reserves of fossil fuels are sufficient to produce nearly a 200%
increase in the carbon dioxide content of the atmosphere.[29]

The report made a number of specific claims about the scientific evidence and implications:

Pollutants *have altered* on a global scale the carbon dioxide
content of the air.[30]

... the data show, *clearly and conclusively*, that from 1958–
1963, the carbon dioxide content of the atmosphere increased
by 1.36%.[31]

We can conclude with fair assurance that at the present time
fossil fuels are the *only source* of CO_2 being added to the ocean-
atmosphere-biosphere system.[32]

By the year 2000 the increase in atmospheric CO_2 will be close
to 25%. This may be sufficient to produce measurable and perhaps marked changes in climate ... [33]

With a 25% increase in atmospheric CO_2, the average temperature near the earth's surface could increase from 0.6
to 4.0 degrees C, depending on the behavior of the atmospheric water vapor content ... *A doubling of CO_2 in the air,
which would happen if a little more than half of the fossil
fuel reserves were consumed, would have about three times the
effect of a twenty-five percent increase* [i.e., 1.8–12.0°C].[34]

In a special message to Congress, President Johnson drew out
three implications. First, he declared that climate change was
already occurring, and fossil fuel use was a key cause: "This
generation has altered the composition of the atmosphere

on a global scale through ... a steady increase in carbon dioxide from the burning of fossil fuels." Second, he asserted that climate change is global problem with political implications: "Large-scale pollution of air ... is no respecter of political boundaries, and its effects extend far beyond those who cause it." Third, he warned that the serious time-lags implied that early, anticipatory action was needed: "The longer we wait to act, the greater the dangers and the larger the problem."[35]

In summary, by 1965 there was high-level political awareness (and concern) about the existence, magnitude, and timing of the climate threat. This casts significant doubt on the "blissfully ignorant" and "relatively recent phenomena" claims, as well as the 1990 benchmark. Still I do not advocate simply adopting an earlier "threshold" of knowledge. Instead, in my view, a better approach would be to embrace a more sophisticated historical understanding that acknowledges an evolution of awareness over time. This should include *accepting increasing responsibility for decisions not to act*, especially when these involve calculated gambles about the level of risk. One advantage of such an approach would be to diffuse some of the unfortunate political attention placed on disputing alleged "all-or-nothing" scientific thresholds for knowledge (such as the 1990 benchmark).[36]

Aside from history, the ignorance defense also faces theoretical difficulties. To begin with, it is worth distinguishing blame from responsibility. Though we do not usually blame those ignorant of what they do, we often hold them accountable. Hence, showing that blame is inappropriate is insufficient to dismiss past emissions. Moreover, the three presumptive arguments provide reasons for holding the ignorant accountable in this case.

First, consider causal responsibility. If I accidentally break something of yours, we usually think that I have some obligation to fix it, even if I was ignorant that my behavior was dangerous, and perhaps even if I could not have known. It remains true that I broke it, and in many contexts that is sufficient. After all, if I am not to fix it, who will? Even if it is not completely fair that I bear the burden, it is often at least *less* unfair than leaving you to bear it alone.

Interestingly, this may hold even if it was not really me who did it. Suppose my dog escapes into your yard, and digs up your vegetable patch. Suppose I have done everything that could reasonably be expected to prevent this (e.g. installed a perfectly good fence, which she has tunneled underneath). I am not usually to *blame* for the damage to your vegetables. Nevertheless, I am accountable. Even though I have acted perfectly responsibly, she is still my dog, and I have to bear the burdens of her activity. Part of what one takes on when one has a dog is these kinds of responsibilities. Perhaps something similar is true of citizenship.

Second, consider fair access. Suppose that I unwittingly deprive you of your share of something and benefit from doing so. Isn't it natural to think that I should step in to help when the problem is discovered? For example, suppose that everyone in the office chips in to order pizza for lunch. You have to dash out for a meeting, and so leave your slices in the refrigerator. I (having already eaten mine) discover and eat yours because I assume that they must be going spare. You return to find that you now don't have any lunch. Is this simply your problem? Usually, we don't think so. Even though I didn't realize at the time that I was taking your pizza, this does not mean that

I have no special obligations. The fact that I ate your lunch remains morally relevant.

Third, according to the general practice argument, we must consider all of the effects, direct and indirect, of endorsing a system where the ignorant are not held liable for their actions. One concern is the creation of perverse incentives, especially around information. For instance, if I am liable only if I know, then other things being equal it is better for me to remain ignorant. This encourages phenomena such as "turning a blind eye," self-deception, and cultivated ignorance. Similarly, if liability requires public endorsement of the relevant facts, then I have incentives to spread disinformation in order to protect myself from liability.[37] Other things being equal, this seems likely to lead to bad results, especially as a long-run strategy for global environmental affairs. Notably, in the context of rapidly escalating global environmental change, establishing a "blissful ignorance" precedent may well turn out to be one of the worse outcomes of the current policy debacle.

4.4.3 "First Come, First Served"

The second objection to corrective justice emerges from a disanalogy. In the pizza case, you have a clear right to the eaten slices, because you have already paid for them. With emissions, it might be argued that latecomers have no such claim. Perhaps it is simply "first come, first served," and hard luck to the tardy.

In my view, this response is too quick. We must ask what initially justifies policies like "first come, first served." Consider one natural explanation. If a resource initially appears to be unlimited, consumers may simply assume that no issues of allocation arise. Everyone can take whatever they want, with no adverse consequences for others. In this case, the principle is not really "first come, first served" (which implies that the resource is limited, so that some may lose out), but rather "free for all" (which does not). Since it is assumed that there is more than enough for everyone, no principle of allocation is needed.

In such a case, what happens if the assumption that the resource is unlimited turns out to be mistaken, so that "free for all" becomes untenable? Do those who have already consumed large shares have no special responsibility to those who have not and now cannot? Does the original argument for "free for all" justify ignoring the past? I think not. If the parties had considered at the outset the possibility that the resource might turn out to be limited, which allocation principle would have seemed more reasonable and fair: "free for all, with no special responsibility for the early users if the resource turns out to be limited," or "free for all, but with early users liable to extra responsibilities if the assumption of unlimitedness turns out to be mistaken"? Offhand, there seem clear reasons to resist ignoring the past: it makes later users vulnerable in an unnecessary way, and provides a potentially costly incentive to consume early and often if possible. Given this, "first come, first served" looks unmotivated. Why adopt an allocation rule that so thoroughly exempts early users from responsibility?

4.4.4 Dead Emitters

The third objection to past emissions emphasizes that, since significant anthropogenic emissions have been occurring since 1750, many past polluters are now dead. Given this, it is said, "polluter pays" principles no longer apply. Instead, what is being proposed is that polluters' descendants pay, because they have benefited from past pollution (usually because of industrialization in their countries). However, "beneficiary pays" principles are unjust. They hold current individuals responsible for emissions that they did not cause, could not have prevented, and in ways that diminish their own opportunities.

Much could be said here, but let me make just three points. First, it is not self-evident that the polluters are dead. For one thing, half of cumulative emissions since 1750 have occurred since 1970. For another, Polluter pays approaches typically refer not to individuals as such but to some entity to which they are connected (e.g., a country, people, or corporation), and these usually have persisted over the time period envisioned. In response, proponents of the dead emitters objection typically argue against the moral relevance of states, by endorsing a strong individualism that claims that only individuals should matter ultimately from the moral point of view. The Chicago lawyers are particularly strident, complaining that the collective responsibility implied by polluter pays has been "rejected by mainstream philosophers as well as institutions such as criminal and tort law" in favor of "the standard assumption of individualism."[38] It is this individualism, rather than dead emitters as such, that give rise to the challenge.

Still, the implicit individualism of the dead emitters argument is more controversial than it initially appears, and often employed in a *highly selective* fashion. On the one hand, collective responsibility is not so easily dismissed. Many international practices rely on it. For example, they assume that the current generation of Americans is liable

for debts incurred by its predecessors, that future generations will accept responsibility for our debts, that we and
they will abide by treaties that our predecessors signed,
and so on. Indeed, it is a challenge to imagine what international relations—or intra-national relations—would
look like if this were not so. Consequently, philosophical
individualists typically concede that states play important
roles in representing individuals and discharging their
moral responsibilities. Given this, we must beware of moral
corruption: e.g., insisting on strong individualism about
climate responsibilities might involve highly selective
attention to how the world usually works.

On the other hand, a very robust individualism would
call into question practices surrounding inherited rights, as
well as inherited responsibilities. Put most baldly, if (say)
Americans are not responsible for at least some of the debts
incurred by their ancestors, *why are we entitled to inherit the
benefits* of those activities? In particular, if we disavow past
American emissions, must we also relinquish inherited
American assets, such as the territory and infrastructure
left to us?

Some radical individualists will not flinch from these
implications. They may be right, especially as a matter of
ideal theory. Importantly, it may also be true that many
vulnerable nations would be willing to give up on claims
of corrective justice *in exchange* for the implementation
of such radical cosmopolitan ideals. For instance, perhaps
Bangladesh and Haiti would be willing to relinquish their
claims to climate justice so long as they were given a fair
(cosmopolitan) share of American and Chinese wealth.
From an ambitious, idealistic point of view, this may even
be a promising approach.

I do not know whether the Chicago lawyers would be willing to push their individualism this far. I would say that it sits awkwardly with their insistence on political feasibility. From my point of view, if the "feasibility" constraint of accepting the existence of nation-states means anything much, it involves accepting some traditional norms of collective responsibility, at least for a while, as part of an ethics of the transition. However, if this is right, it is not clear why past emissions are off the table.

4.4.5 Resource Isolationism

A final objection returns us to the issue of isolation. Even if past injustice is relevant, isn't it *arbitrary* to single out carbon emissions from other global resources (e.g., land, oil, coal, uranium, gold, diamonds), and from the general legacy of international wrongdoing (e.g., the slave trade, colonialism)?[39] Simon Caney, for instance, vigorously objects to the "method of isolation," insisting that "someone who is committed to equality of commonly held natural resources should embrace a principle granting everyone an equal share of the total value of all these global resources combined."[40]

I agree that in some ways this is a serious challenge. It affects not only ideal theory, but also some forms of the ethics of the transition. One way to respond would be to embrace an all-inclusive approach. For example, cosmopolitan individualists often propose solutions to the general resource problem (such as a global tax on natural resources, or a global resources dividend) that might form part of an ethics of the transition.[41] Whether a more isolationist approach is a better strategy remains an open question.

Still, there are grounds for treating climate separately. First, pragmatically, the current state system explicitly regards a large number of these resource justice questions (e.g., oil, gold) as largely settled, and because of the principle of territorial rights. However, the atmosphere is not like this. It is no one's territory, and until recently people assumed that its capacity to absorb greenhouse gases was unlimited and so raised no issues of fairness in distribution ("free for all"). Unlike the other "resources," then, what to say about this one—as a matter of international law and politics—is very much up for grabs. Moreover, what is said has major implications for other global "resources" as human impact grows.

Second, treating climate differently may make theoretical sense. Specifically, I suspect that it is a mistake to view the atmosphere as a "resource" at all, and especially a resource for "dumping things into." For instance, a better way to look at our natural environment might be as a background condition against which we live and other things can *count* as resources. If so, the atmosphere is not really akin to gold or oil. Its role is deeper. One aspect, of course, is basic life support; but there are others, including serving as part of the fabric of the place where we live, and which partly determines who we are and what we can do.

Such thoughts might serve as the seeds of a more robust philosophical position in climate ethics. For instance, they may explain part of the extortion claim: why holding the climate hostage is not just a matter of bare resources, or mere money. This is not the place to fully develop such an ethics.[42] Still, I will pursue the issue just a little, to motivate an anti-extortion approach. To do so, let us turn to the trajectory question.

4.5 TRAJECTORY

Conventional climate policy implicitly involves envisioning a long-term aim, and then deciding how quickly to achieve that aim.

4.5.1 Feasibility Revisited

Numerous targets have been proposed. Some seek to limit global temperature rise (e.g., to 2°C), others aim at a specific atmospheric concentration (e.g., 350, 450, or 550 ppm of CO_2 equivalent), and still others propose not exceeding a given total of human emissions (e.g., one trillion tons).

The differences between these targets are not much discussed. One reason is presumably that, since all the candidates are far from "business as usual" projections, advocates assume that a move towards any one is a move substantially in the right direction; hence they are disinclined to highlight disagreements. A second reason is that there is substantial consensus on the speed at which humanity should reach these long-term goals (e.g., 50%-80% emissions cuts by 2050).

This political consensus is encouraging. Nevertheless, quantitative targets tend to obscure underlying ethical issues. In particular, although much talk of specific percentage reductions is carried out in the language of "feasibility," and so seems technical, this is a mistake. Presumably, it is perfectly *technically feasible* for us all to reduce our emissions by 50%-80% tomorrow, or even eliminate them. We could, after all, just turn off our electricity, refuse to drive, and so on. The problem is not that this cannot be done; it is rather that the implications are bleak. Given our current infrastructure, we assume that a very rapid reduction would cause social and economic chaos, including humanitarian disaster and severe dislocation, for the current generation. If this is correct, we are justified in dismissing such drastic measures. However, the justification is ethical: a policy that demanded them of us would be profoundly unjust.

Moving away from "feasibility" makes a difference. Even if any emissions cuts would be disruptive to some extent, presumably at some point the risks imposed on future generations are severe enough to outweigh them. Perhaps current proposals—such as 20% by 2020—capture the appropriate tradeoff point. Nevertheless, it would be nice to see some argument for this claim, especially since an issue of intergenerational justice is at stake, and we are likely—given the perfect moral storm—to be biased in our own favor.

Given the theoretical storm, robust, "shovel ready" solutions are unavailable. I propose that we take the elephant in the room—the threat of intergenerational extortion—as a starting point, for both an ethics of the transition and for making progress on ideal theory. With this in mind, let me gesture at three key ideas. Though each requires much further development, together they suggest at least one way forward for climate ethics.

4.5.2 Self-Defense

The first idea amounts to an error theory. Some of the rhetorical appeal of the economic realist position rests on a misdiagnosis. The essential rationale for the current generation's continuing with relatively high levels of emissions in the near term is one of self-defense, rather than self-interest more generally. Given the tightening emissions budget, and strong evidence of severe risks to the future, the initial default answer to the trajectory question is that we should cease emitting. At first glance, the only even vaguely adequate reason for not doing so is that such a demand is unreasonable, in the sense of violating fundamental entitlements that are protected by a right to self-defense.

To see the appeal of this view, imagine that our generation faced a different intergenerational issue. Suppose the normal food supply became contaminated with a pesticide that has only intergenerational effects. Specifically, the pesticide caused increasingly severe and painful deformities three generations hence and then for ten more generations, in direct proportion to the amount used by us. What could justify our continuing to employ it?

The obvious answer is "very little." Among the more dubious possibilities would be claims such as "food produced this way is cheaper than organic," or "the future will be richer if we keep using the pesticide," or "the future should pay us to stop using the pesticide." More promising answers would include "we will starve," or "our families will fall into poverty," or "given the structure of our economy, our community will collapse." These all involve severe consequences for the agent, and so appeal to something like a right to self-defense.

Still, the self-defense framing does not leave everything as it was. On the one hand, such a right would come with *sharp limits*. Consider some examples. First, one is normally required to use other, nonharmful means of escaping the threat if possible (e.g., in the pesticide case, if you can afford organic food without sacrificing something even

vaguely comparable to what is inflicted on future generations, buy it). Second, if no nonharmful means is immediately available but the threat is ongoing, one may harm, but
there remains a strong obligation to find a nonharmful way
out as quickly as possible, even if this involves costs to oneself. (For example, if an immediate pesticide ban will cause
mass unemployment and collapse in some communities,
work hard to phase it out quickly, or to find new jobs or
even new communities for the displaced.) Third, even when
harm is permitted, one is allowed only to use the minimum
force necessary (e.g., the minimum amount of pesticide
that is absolutely necessary, or the amount that minimizes
total exposure for future people). Fourth, one must provide compensation for the unavoided harm (e.g., invest in
research to find a cure, medical infrastructure to care for
victims, etc.).

On the other hand, the right to self-defense is *not necessarily universally applicable, or even welcomed.* First, according to some, the right does not apply if one is deemed the
aggressor (e.g., an extortionist). Second, some believe that
self-defense is impermissible if the victim is innocent.
Third, and perhaps most interestingly, as self-defense is a
right rather than a duty or obligation, some may choose not
to invoke it, at least not as strenuously as is permissible.
For instance, they may choose to take on extra burdens
themselves rather than pass them on to their children or
grandchildren; or they may simply prefer, as a moral matter, not to inflict extra burdens on others, even if they have
a right to do so. In my view, these are important but under-
appreciated elements of climate ethics.

The self-defense approach helps to explain the appeal
of some versions of economic realism. Specifically, it can

account for the thoughts that self-interest suggests some kind of constraint on what can legitimately be asked of people, and that nevertheless we need to phase out emissions quickly. It does so without suggesting that ethics is irrelevant or counterproductive, and without endorsing extortion. Nevertheless, it involves stringent restrictions, with profound implications for climate policy.

4.5.3 Basic Physical Structure

The second key idea considers the perspective of victims of climate extortion. Extortion involves the use of an asymmetry of power by the more powerful actor. Still, this is not enough to make it morally objectionable. People regularly profit from superior bargaining positions because (for example) they are better qualified, or produce better products. The problem is not benefiting from asymmetry as such, but rather from specific asymmetries under certain conditions. For instance, many classic extortion cases involve extortionists benefiting from threats that they will deploy their superior capacity for physical violence. Yet when the Mafia threatens your daughter, we do not think that this capacity is a "resource" of theirs that they are morally entitled to use as a basis for demanding benefits. Similarly, there seems something deeply wrong with the current generation imposing climate burdens on future generations, and this is not legitimized by the bare fact that time's arrow gives them the power to do so.

How might ethics pursue this thought? One avenue is to highlight the imposition of serious risks of death and suffering on innocent future people. These seem closely analogous to threats of physical violence, and are (rightly) cited by most climate ethicists. Still, (as signaled above) I suspect climate plays a deeper role.[43] Consider an (admittedly imperfect) analogy. One of John Rawls' most influential ideas is that questions of justice arise in contemporary societies in part because they profoundly *shape the basic life prospects* of their citizens, through their creation of a set of powerful institutions backed by coercive force that constitute the "basic structure" of society. Similarly, one might understand anthropogenic climate change as threatening what we might call the *basic physical structure* of the planet. Serious interference with the climate system would shape the lives of very many people around the world, and especially future generations, deeply and pervasively, to the extent of becoming at least a major determinant of their life prospects, and perhaps the dominant factor. Arguably, climate extortion exploits this fact, and this is a central part of what makes it ethically problematic.

This thought is perhaps best illustrated by the recent proposals to "geoengineer" the planet by injecting sulfates into the stratosphere to reduce incoming sunlight, and thereby cool down the surface. Alongside (serious) feasibility concerns, one major worry is that those who intentionally aim to take control of the climate system attempt to master the basic physical structure of the planet, making choices that shape the basic life prospects of everyone. Given this, a central question should be, "What would would-be climate controllers *owe* those put under their yoke, especially in terms of procedural and distributive justice?" Arguably, this question raises profound issues of global politics and political philosophy, and ones that the existing debate does not take seriously enough. For instance, (as an illustration) in the case of the domestic basic structure, Rawlsians tend to think that what is owed is extensive (e.g., the infamous *difference principle* demands that social and economic inequalities should be arranged so as to maximally benefit the least well off.) Yet in the early debate about geoengineering, there is little discussion of strong norms being adopted. Indeed, the contrary assumption is common: some believe that sulfate injection is politically easier than reducing emissions because one country (or even a rich individual) could practice it without the cooperation of others. In my view, such attitudes involve a serious ethical mistake: taking over the basic physical structure involves profound political issues, including of basic justice.

4.5.4 Limits

My third key idea rebels against the profound politicization of nature suggested by both sulfate injection and climate extortion. We should resist turning global climate into a political domain where justice and legitimacy are the salient values. To motivate this, consider another analogy:

Suppose scientists develop a device that allows someone to insert thoughts inside another's brain. Few would argue that the main concern of public policy should be to facilitate the most efficient methods of thought control. Neither would many suggest that the most important questions would concern what thought controllers owe their victims in terms of procedural justice and the distribution of burdens (however demanding). Instead, thought control ought simply to be off limits. Governments should prevent such power being exerted (just as it should prevent the kidnapping of people's daughters).

Many of us have the same instincts when it comes to climate. We are appalled to get to the point where sulfate injection schemes are even on the table, and shocked by the suggestion that we should load future generations with debt so that they can "bribe" us to stop. We do not want a grand ethics that tells us how to do "just" or "efficient" geo-engineering (for example), any more than we want plans for "just" or "efficient" kidnapping or thought control. We favor *withdrawing* interference from the climate (e.g., mitigation) over "managing" it. In short, we seek an ethics that forecloses some options as incompatible with who we are and aspire to be. One sign of this is that even if we get to the point where geoengineering becomes inevitable, this saddens us and comes at some further moral cost.[44]

As with thought control, the central issues involve (first) concerns about unnecessarily extending the powers of humans over one another, and (second) a broader vision of how we want to live. In the climate case, "how we want to live" includes not just our relationships with each other, here and now, but also with future generations and nature. It includes taking seriously the question of what limits we want to set in order to respect those relationships. Without limits, we give free rein to extortion, and risk becoming "the scum of the Earth."

4.6 A GLOBAL CONSTITUTIONAL CONVENTION

My three key ideas offer only the seeds from which one version of climate ethics might grow, for the purposes of illustration. Robust theories remain far away. In the meantime, we must do what we can in the ethics of the transition, including making suggestions about where to go in the absence of such theories. I have argued that there is enough ethical convergence to provide guidelines for "rough justice" moving forward, at least in the initial steps. Importantly, current behavior, especially among the affluent in the richer countries, is so far from meeting any reasonable ethical standards that we cannot plausibly complain that the theoretical storm is our most serious problem right now.

I close with one further, quasi-pragmatic suggestion. In my view, our biggest problem involves an institutional gap, especially when it comes to intergenerational concern. The most obvious move is to close that gap. Elsewhere, I argue that the natural strategy involves calling for a global constitutional convention (akin to the US constitutional convention of 1787) focused on future generations, with a mandate to confront the ethics of the transition head on, in a way that makes the issues visible to all.[45]

Such a proposal directly confronts the short-termism and narrow economic focus of current, and especially national, institutions. Consequently, it will seem profoundly "infeasible" to some. Still I believe that some such confrontation is unavoidable. Twenty-five years of political horse trading between conventional institutions shows no sign of protecting future people, or the planet more generally. If we are to avoid climate extortion, we need a more ambitious vision of "what works."

Moreover, the convention proposal remains comparatively modest in light of the alternatives. At its heart, the idea is simply that the world's peoples must discuss the current impasse at a level that takes seriously the problem we face. In addition, we should not be overly pessimistic. A constitutional convention may be something that not just global publics, but also leaders of current institutions—knowing the institutional limitations better than most—can bring themselves to support. If we have genuine intergenerational concern, the current situation should bother us all deeply. Moreover, the call for a global constitutional convention aimed at protecting the future reflects a noble ambition. If our generation can fix the current institutional gap, we can prepare humanity for the future in a way that makes us not just "green," but genuinely great. Isn't it worth a try?

Notes

1. Jonathan Pickering, Steve Vanderheiden, and Seumas Miller, "If Equity's In, We're Out," *Ethics & International Affairs* 26, no.4 (2012): 423–443.
2. John Rawls, *A Theory of Justice*, Rev. Ed. (Oxford: Oxford University Press, 1999).
3. Eric A. Posner and Cass R. Sunstein, "Pay China to Cut Emissions," *The Financial Times*, August 5, 2007, http://www.ft.com/intl/cms/s/0/e67a8166-436d-11dc-a065-0000779fd2ac.html#axzz39YR7J6pj.
4. Eric A. Posner and David A. Weisbach, *Climate Change Justice* (Princeton, NJ: Princeton University Press, 2010), 86.
5. Eric A. Posner and C.R. Sunstein, "Climate Change Justice," *Georgetown Law Journal* 96, no. 5 (2008), 1569.
6. Posner and Sunstein, "Pay China to Cut Emissions."
7. Peter Singer, "One Atmosphere," in *Climate Ethics: Essential Readings*, ed. Stephen M. Gardiner, et al. (Oxford: Oxford University Press, 2010), 181–199.
8. John Broome, *Climate Matters* (New York: Norton, 2012). See also Matthew Rendall, "Climate Change and the Threat of Disaster: The Moral Case for Taking out Insurance at Our Grandchildren's Expense," *Political Studies* 59, no. 4 (2011): 884–899.
9. Broome, *Climate Matters*, 45–48.

10. I do not assume that future people ought never to share in the costs of climate action. For instance, once a generation has been handed a significant burden through the misbehavior of its predecessors, some sharing (of *that* burden) might be appropriate. As the perfect storm rolls on to produce second- and third-generation climate ethics, there are further nuances.

11. Rawls, *A Theory of Justice: Revised Edition*, 3–4.

12. Singer, "One Atmosphere."

13. International Energy Agency, "World Energy Outlook," Paris: International Energy Agency, 2009, https://www.iea.org/textbase/npsum/weo2009sum.pdf.

14. T.A. Boden, G. Marland, and R.J. Andres, "Global, Regional, and National Fossil-Fuel CO_2 Emissions," Global, Regional, and National Annual Time Series, Oak Ridge, Tennesee: Carbon Dioxide Information Analysis Center, Oak Ridge National Laboratory, U.S. Department of Energy, 2009, http://cdiac.ornl.gov/trends/emis/overview.html.

15. World Bank, "CO_2 Emissions: Metric Tons per Capita," 2014, http://data.worldbank.org/indicator/EN.ATM.CO2E.PC.

16. International Monetary Fund, "World Economic Outlook Database, April 2014," 2014, http://www.imf.org/external/pubs/ft/weo/2014/01/weodata/weorept.aspx.

17. World Bank, "Overview," 2015, http://www.worldbank.org/en/topic/poverty/overview.

18. Singer, "One Atmosphere," 193–194 in Gardiner, et al. 2010.

19. For attempts to integrate such concerns, see Paul Baer, Tom Athanasiou, Sivan Kartha, et al. *The Greenhouse Gas Development Rights Framework: The Right to Development in a Climate Constrained World* (Berlin: Berlin Heinrich Böll Foundation, 2007); Shoibal Chakravarty, et al., "Sharing Global CO2 Emission Reductions among One Billion High Emitters," *Proceedings of the National Academy of Sciences of the United States of America* 106, no. 29 (2009): 11884–11888.

20. Singer, "One Atmosphere," 185.

21. United States Energy Information Administration, "International Energy Statistics," 2014, http://www.eia.gov/cfapps/ipdbproject/IEDIndex3.cfm?tid=90&pid=44&aid=8.

22. Posner and Weisbach, *Climate Change Justice*, 86.

23. Thomas Schelling, "The Cost of Combating Global Warming: Facing the Tradeoffs," *Foreign Affairs* 76, no. 6 (1997): 11.

24. Stephen M. Gardiner, *A Perfect Moral Storm* (New York: Oxford University Press, 2011), chap. 8.

25. Schelling, "The Cost of Combating Global Warming: Facing the Tradeoffs," 10.

26. Posner and Weisbach, *Climate Change Justice*, 74.

27. Ibid., 73.

28. Darren Samuelsohn, "No 'Pass' for Developing Countries in Next Climate Treaty, Says U.S. Envoy," *The New York Times*, December 9, 2009, http://www.nytimes.com/gwire/2009/12/09/09greenwire-no-pass-for-developing-countries-in-next-clima-98557.html?pagewanted=all.

29. Environmental Pollution Panel, *Restoring the Quality of Our Environment*, Report of the Environmental Pollution Panel, President's Science Advisory Committee, Washington, DC: The White House, November 1965, 126, http://dge.stanford.edu/labs/caldeiralab/Caldeira%20downloads/PSAC,%201965,%20Restoring%20the%20Quality%20of%20Our%20Environment.pdf.

30. Ibid., 1 (emphasis added).

31. Ibid., 116 (emphasis added).

32. Ibid., 119 (emphasis added).

33. Ibid., 126–127.

34. Ibid., 121 (insert added).

35. President Lyndon B. Johnson, "Special Message to the Congress on Conservation and Restoration of Natural Beauty," February 8, 1965, Santa Barbara, CA: The American Presidency Project, 2015, http://www.lbjlib.utexas.edu/johnson/archives.hom/speeches.hom/650208.asp.

36. Stephen M. Gardiner, "Justice and the Simple Threshold View of Past Emissions," unpublished.

37. E.M. Conway and Naomi Oreskes, *Merchants of Doubt* (New York: Bloomsbury, 2010).

38. Posner and Weisbach, *Climate Change Justice*, 101.

39. For example, see Simon Caney, "Justice and the Distribution of Greenhouse Gas Emissions," *Journal of Global Ethics* 5, no. 2 (2009): 125–146; Simon Caney, "Just Emissions," *Philosophy & Public Affairs* 40, no. 4 (2012): 255–300; Megan Blomfield, "Global Common Resources and the Just Distribution of Emission Shares," *Journal of Political Philosophy* 21, no. 3 (2013): 283–304; Megan Blomfield, "Climate Change and the Moral Significance of Historical Injustice in Natural Resource Governance," in *The Ethics of Climate Governance*, eds. Aaron Maltais and Catriona McKinnon (London: Rowman and Littlefield International, 2015).

40. Simon Caney, "Just Emissions," 271–291.

41. Thomas Pogge, *World Poverty and Human Rights: Cosmopolitan Responsibilities and Reforms* (Cambridge: Polity, 2002); Charles Beitz, "Justice and International Relations," *Philosophy & Public Affairs* 4, no. 4 (1975): 360–389.

42. Stephen M. Gardiner, "Rawls and Climate Change: Does Rawlsian Political Philosophy Pass the Global Test?" in "Climate Change and Liberal Priorites," eds. Gideon Calder and Catriona McKinnon, special issue, *Critical Review of International Social and Political Philosophy* 14, no. 2 (2011): 125–151. doi:10.1080/13698230.2011.529705.

43. This section draws on Stephen M. Gardiner, "Geoengineering: Ethical Questions for Deliberate Climate Manipulators," in *The Oxford Handbook on Environmental Ethics*, eds. Stephen Gardiner and Allen Thompson (Oxford University Press: Oxford, in press).

44. Gardiner, *A Perfect Moral Storm*, ch. 10.

45. Stephen M. Gardiner, "A Call for a Global Constitutional Convention Focused on Future Generations," *Ethics and International Affairs* 28, no. 3 (2014): 299–315.

PART II

THE PROBLEMS WITH

CLIMATE ETHICS BY DAVID

A. WEISBACH

5

Introduction to Part II

ACTIVITIES YOU AND I engage in every day without thought—heating and cooling our homes, turning on the lights, taking a warm shower, commuting, eating—harm other people. These activities require energy.[1] Most of our energy comes from fossil fuels. Using fossil fuels results in emissions of carbon dioxide, which will cause climate change. The harms from climate change will range from mild to possibly catastrophic. Many people's livelihood, food supply, or place of living will be altered or destroyed. And many of the people causing the harms are wealthy or live in wealthy nations. Many of the victims will be poor.

Notwithstanding these harms, and notwithstanding more than twenty years of international negotiations to establish limits, emissions of greenhouse gases continue to rise. Since the first major climate treaty, the 1992 Framework Convention on Climate Change, annual emissions have gone up by more than a third.[2] They show no sign of slowing. Developed nations overall have stabilized emissions but have not substantially reduced them. Fast developing nations have rapidly increased their energy use. As a result, their emissions have doubled in the period from 1992 to today. Extensive negotiations have managed to produce agreements to agree sometime down the road.

How should ethics help us evaluate these facts and help us to decide what to do? One view is that climate change is primarily an ethical problem. Climate change seems to raise many questions of ethics or justice, such as what are our duties to people who live in other countries or who live in the future, what is the just distribution of wealth, how should we divide global resources, and when, if ever, is it permissible to harm others. Under this view, ethics can tell you whether you can take that warm shower and have a cup of coffee tomorrow morning.

Many people take the view that climate change is primarily an ethical problem. My co-author and debating partner, Steven Gardiner, states simply "Climate change is an ethical issue."[3] It is, moreover, not sufficient to consider our own self-interest to determine how to address climate change. Gardiner argues:

> The dominant reason for acting on climate change is not that it would make us better off. It is that not acting involves taking advantage of the poor, the future, and nature. We can hope that refraining from such exploitation is good (or at least not bad) for us, especially in terms of current lifestyles and those to which we aspire. But such hope is and should not be our primary ground for acting. After all, morally speaking, *we must act in any case.*[4]

Philosophers often go on to argue that climate policy has to take a particular form to meet ethical constraints. That is, not only does ethics help determine the values we should use in designing climate policy. It tells us which specific climate change policies are the right ones. Philosophers argue, for example, that climate policy must be designed to address distributive concerns, that climate policy must take into account past injustices, and that climate change policy must rely on ethics to determine who has the right to emit how much in the future.

For example, the philosopher Peter Singer argues that total emissions of greenhouse gases should be capped at one trillion tons of carbon, and that rights to emit up to this cap should be divided among countries on a per capita basis using population projections for the year 2050.[5] The underlying theory is based on a theory of equality: all humans have an equal right to the atmosphere so allocating use on a per capita basis is required by justice.

Others have tried to find a numerical basis for allocating responsibility for past emissions. These calculations are motivated by notions of duties to avoid harming others and a corresponding obligation to compensate for any harm done in the past. Teams of scientists, in an attempt to support the philosophical arguments, have performed elaborate calculations of who owes whom what as if they were testifying as expert witnesses on damage valuations in a trial.[6]

Other philosophers are not quite this specific but still argue that the shape of climate policy should reflect particular ethical concerns. For example, many argue that to address distributive concerns, wealthy nations should have greater duties to reduce emissions than poor nations do.[7] The Kyoto Protocol, as I write, the only climate treaty that imposes binding obligations to reduce emissions, reflects distributive concerns. It imposes binding caps on wealthy countries and no obligations whatsoever on anyone else.

These philosophers are mistaken. Mainstream philosophical claims about climate change suffer from serious and systematic flaws.[8] The first is what I will call climate change blinders. We wear climate change blinders when we think that climate change policy is the tool we need to use to solve whatever ethical problem happens to be in front of us, forgetting that there are many tools and policies at our disposal.

An example is the widely held belief that climate change policy should be designed with distributive goals in mind. There are vast differences in wealth around the globe. People argue that climate policy should be designed to address this ethical problem. As a result, they recommend against choosing the lowest cost method of reducing emissions. Instead, wealthier nations should reduce more even if the costs of their reductions are higher than the costs of reductions found elsewhere. Wealthy nations can, after all, better afford these costs. Claims of this sort have led to policies such as the Kyoto Protocol in which wealthy nations have a binding obligation to reduce emissions while developing nations have no obligations whatsoever.[9]

There are, however, many ways to achieve distributive goals other than through a climate treaty. We might, for example, change trade policies, remove subsidies for domestic industries that hurt competing industries in developing countries, change the rules governing intellectual property, or provide free goods of various sorts—such as mosquito nets, vaccines, or chlorine to sterilize water. Perhaps helping to develop better governance structures or reducing corruption is the best way to help the poor. Microcredit, women's rights, deworming, or indoor plumbing all might be central to that goal. We don't know which, if any, of these policies is best. A great number of people have devoted their lives to finding solutions. Many approaches have been tried. Few have worked.

Philosophy can help us understand the value of improving the distribution of well-being, but it does not tell us that climate change policy is the best way or the necessary way to achieve this goal. By trying to redistribute within a climate change policy, we may be choosing a way of helping the poor that is more expensive and less effective than other forms of meeting distributive obligations. Moreover, by trying to redistribute within a climate change policy, we risk failing to achieve core climate change goals. Climate change blinders prevent us from seeing that other tools are available and that we should pick the combination of policies that best achieves our multiple goals. By focusing on climate change and ignoring the broader policy context, the arguments risk producing policies that fail to achieve any of their stated goals in an effective manner.

The second flaw with many ethical arguments about climate change is that they often produce recommendations which violate basic feasibility constraints. An example is the commonly made proposal to divide the atmosphere into equal portions and to distribute those portions in the form of rights to emit carbon dioxide equally among the world's population—creating equal per capita emissions rights. Singer's proposal, mentioned above, is an example. The approach seems ethically appealing because it treats all humans equally. The atmosphere, some say, is a common resource of all humans. Allocating the right to use it equally is, one might think, a minimal requirement of justice. An American should not have the right to emit more than a Brazilian or Bangladeshi because of the mere happenstance of birth.

This approach, however, is completely infeasible. The size of the immediate wealth transfers would exceed the transfers that are currently made from rich to poor countries by several orders of magnitude. If done in the manner that analysts propose—hand out tradable permits to people or countries and let them sell them for cash—the transfers would be done without any sort of controls on the use of the money, controls which are now and have always been required. Yet philosophers ignore such problems. Peter Singer, a proponent of this approach, goes so far as to argue that the equal per capita approach is suitable as a political compromise without attempting to give a single historical example of voluntary transfers of this magnitude. The policy recommendation is utopian in the bad sense of the word.

Perhaps Singer and others take the view that we should determine what is right entirely without regard to whether it is feasible. If ethics requires men to behave like angels, that is what is required even if nobody can meet that goal. If the wealthy countries have to give vast sums to other countries as part of a climate treaty to be ethical, it does not matter that they would never do this. It is still required. Failing to meet ethical demands does not diminish the demands.

Another view is that philosophy should not stray too far, if at all, beyond feasibility constraints if it is to be helpful in telling us what to do. Climate change is an urgent problem. We need to find solutions that will work. Claims based on ethics that ignore basic feasibility constraints are at best idle chatter, and at worst, divert our attention from actual solutions.

The claim that ethics should be limited by feasibility, taken too far, chafes. Ethics tells us what we ought to do. It is, by its nature, demanding. If we can just refuse to do what is required by ethics, claiming it is infeasible, ethics is empty. I won't stop my fist from moving in the direction of your nose. Therefore, I have no ethical obligation not to assault you.

A serious look at how ethics and ethically informed policies operate, however, shows that feasibility is central.[10] Consider the design of a progressive tax system. A progressive tax system improves the distribution of income, reducing inequality and improving welfare. It is ethically desirable. To pursue this goal, we could design a tax system that ignores feasibility constraints in which people are told to reveal their income and pay taxes on that income without audits or sanctions and without considering the effects of the tax system on incentives to work and save. Ideally, tax rates on high levels of income would be at or near 100%, paid voluntarily. Ethics must be demanding. But of course, nobody would comply. The system is infeasible and would not improve anyone's welfare. Designing a tax system that pursues the ethical goal of improving the distribution of resources requires understanding what is feasible and what is not. Designing policies that comply with feasibility constraints produces better, more ethical results, than ignoring them and making idle demands.

Philosophers outside of the climate change context understand this, and use feasibility constraints when considering the design of political institutions. For example, John Rawls, in *Political Liberalism*, argues that a conception of justice must be capable of forming the basis of a stable society.[11] A stable society is one that is "willingly and freely supported by at least a substantial majority of its politically active citizens." Thomas Nagel argues that "the motivations that are morally required of us must be practically and psychologically possible, otherwise our political theory will be utopian in the bad sense.[12] Elizabeth Anderson forcefully argues for affirmative action because of the infeasibility of a truly color-blind society.[13]

How tight the feasibility limits should be will depend on the context and there are no hard and fast rules. If ethics is not demanding at all, it can achieve nothing, but if it is utopian, it is not helpful in designing policies, policies that are urgently needed in the case of climate change.

Nations are the principle actors in climate change and feasibility constraints have particular force when we are considering actions taken by nations. States have not historically been willing to enter into treaties that they expect will make them substantially worse off so that they can help other nations or people.[14] Treaties are the result of negotiations between governments trying to obtain particular benefits through mutual exchange.

This is self-evidently true about run-of-the-mill treaties such as treaties governing taxes, embassies, communication standards, overflight agreements, and the like. It is also true for the closest analogy to a climate treaty, the Montreal Protocol, where nations agreed to reduce the emissions of ozone depleting chemicals to limit the size of the ozone hole. The United States approached Europe for treaty negotiations only after the United States determined that unilateral action would be in its interest.[15] Developing nations are paid to comply, so it was in their self-interest as well.

Even human rights treaties are not an exception to the pursuit of national self-interest. These treaties did not require liberal democracies to change their behavior either because they were already in compliance or, where the treaties might have required a change in behavior, nations issued reservations to excuse them from taking on additional obligations. Authoritarian states entered into the treaties in return for bribes or in response to threats. They mostly ignored the treaties, regardless, so there were benefits and no cost. Transitional states may be the only exception: some of them might have changed their behavior in response to human rights treaties in a way that might seem to be against their interest. According to the empirical research, however, those states entered into human rights treaties because liberal governments then in power wanted to lock in liberal rights.[16] The treaties were in the perceived self-interest of the signatory governments.

Nations sometimes make mistakes and end up with treaties that harm rather than benefit them. Not every nation always achieves its own desired end. Nevertheless, nations consistently pursue their perceived self-interest. Ethical arguments that ask wealthy nations to enter into treaties that systematically make them worse off violate basic feasibility constraints. They are akin to designing an "ethical" tax system based on people voluntarily handing over vast portions of their wealth. Better results will be achieved by paying attention to feasibility.

To establish these claims, chapter 7 will analyze three types of arguments from ethics that have been made regarding climate policy: claims based on theories of distributive justice which conclude that climate policy needs to take the distribution of income, wealth, or other items of value, into account; claims based on theories of corrective justice which conclude that climate policy needs to be adjusted for past wrongful emissions; and claims based on theories of equality which conclude that rights to emit carbon dioxide need to be allocated equally to all people. I will show that each of these theories suffers from climate change blinders and violates basic feasibility constraints.

5.1 SELF-INTEREST VS. ETHICS

In the quote taken from Stephen Gardiner's essay, he contrasts approaches to climate change based on ethics with approaches based on self-interest. Gardiner separately argues that self-interest is "at best a side issue and at worst just another vehicle for procrastination and moral corruption."[17] To help understand the force of this view, it is worth examining what it would mean to pursue our self-interest with respect to climate policy. Does it lead to moral corruption?

Self-interest, I will argue in chapter 6, demands that we pursue aggressive policies to reduce emissions, policies that are far more ambitious than those currently on the table. The core reason is that current emissions levels risk causing terrible harms in the relatively near future, harms that will affect us, our children, and our grandchildren.[18] To prevent these harms, we need to reduce emissions rapidly. Because of the vast fossil fuel infrastructure, reducing emissions rapidly requires acting now. The United States alone has to replace about $6 trillion of durable infrastructure. This will take time. Even if it is likely that the worst of climate change will not happen because the climate is more stable than we currently think, the risk of very bad outcomes is sufficient to mean that it is in our self-interest to act now.

The worry with following self-interest is that it might be in our self-interest to pollute, leaving others to bear the consequences. If I can be better off by polluting, why should I care that you are worse off, at least if I just follow pure self-interest? Perhaps we need to invoke moral or ethical concerns for others to address the problem.

Self-interest, however, need not support this sort of behavior. To see why, suppose we live around a plot of land that everyone can use without charge. Because it is free to use, everyone throws their garbage on the open land, and we expect that soon the land will be polluted and unusable, creating a great loss.

It is in our self-interest to find a way to govern the use of the land. Ungoverned, the common land is tragically wasted. Governed, we can maximize its value by determining whether and how we use it. We might charge people to use it so that they take the costs into account when making choices. We might impose caps on its use. We might come up with a market-based mechanism to determine use. Or we might provide explicit rules for how it is to be used and by whom. Regardless, it is in our self-interest to prevent overuse because doing so preserves the value of the land, making us better off.

Finding a way to limit the use of the land might require an agreement of the people in the community or if there is a government, a decision by the government that reflects and enforces the will of the community.[19] It might require rules and regulations. Self-interest may not mean free markets in the sense of no controls on pollution.

Reaching an agreement may be difficult. People who currently throw their garbage on the land for free might object to limits. They are used to using the land for free and might view free use as a right. And they might be powerful enough to prevent community action. In some real world cases, resources were wasted because of overuse. Jared Diamond's book *Collapse: How Societies Choose to Fail or Succeed* details cases where societies failed to govern their resources wisely.[20] These societies ultimately failed, creating a great loss.

We might say that failing to govern the land, allowing everyone to use it as a garbage dump, is unethical because it hurts other users of the land. Perhaps some people throw in more garbage than others, or someone or their parents threw in more in the past. We might call on ethics to tell us what to do with the common resource.

Better governance of the land, including rules that prohibit free dumping, however, can arise solely as a result of the pursuit of self-interest. It is in our self-interest to manage the land to maximize its value, to prevent it from being wasted. We think it tragic when a society fails because it did not govern its resources wisely. Those societies acted foolishly. They failed to wisely pursue their own ends. It was perhaps unethical but it was also just foolish.

This precise reasoning holds for the use of the atmosphere. Climate change is caused by people dumping their garbage—carbon dioxide and other greenhouse gases—into the atmosphere. Wise use of this vital resource would limit or entirely prohibit this dumping. Wise use is in our self-interest. The argument, moreover, works across space—the people currently situated around the common land or all of us currently using the Earth's atmosphere—and across time—the people alive now and in the future.

Acting in one's self-interest to protect a common resource is not the same as acting selfishly or acting based on narrow and short-term pay-offs. It is not the same as money-grubbing or caring only about the immediate future. It encompasses all sorts of views of the good life. The Greeks called this idea *eudaimonia*, the totality of things one cares about. It encompasses caring for others. Back-to-nature hippies are pursuing their view of the good life and, therefore, self-interest. So are Wall Street bankers. Spending resources to educate my children is in my self-interest because I care about them. But even using narrow notions of self-interest, it is in our self-interest to wisely govern use of the atmosphere, to reduce emissions, starting now, and to do so rapidly. We need to do so to prevent very serious harms to ourselves, our children, and our grandchildren. This is not morally corrupt.

5.2 IS ETHICS INEVITABLE?

Some claim that issues of ethics, justice, and fairness are inevitable in a climate treaty because poor or low-polluting nations will not accept a treaty that treats them unfairly.[21] Poor nations have argued that they should be compensated for past use of the atmosphere by rich nations, that they have a right to use fossil fuels to develop, and that rich nations should shoulder most of the burden of emissions reductions because they are rich. Because poor nations will demand a just treaty and because we cannot limit climate change without the participation of poor nations, issues of justice are, some argue, unavoidable.

The problem with this argument is that it is not the same as an argument that a theory of justice should apply to determine the shape of a climate treaty. Consider the contrast:

Theory of justice applies: There is a theory of justice that determines what nations should do to reduce emissions.

Nations are obligated to consult this theory to determine what to do.

Nations bargain based on perceived notions of fairness: A set of nations bargains for a treaty outcome based on a demand for fair treatment. Other nations may reject this view of fairness and bargain based on other grounds. A treaty is reached on common ground.

The two cases are not the same. In the first case, a theory of justice tells people or nations how they ought to behave. We could read philosophical treatises to determine what we must do. In the second case, nations are simply bargaining based on what they want to get out of the treaty. We could substitute just about any other view about what is desirable in a climate treaty in place of fairness and nothing would change in the second case but the first case would be completely gutted. For example suppose that some nations bargain based on racial hatred, long-simmering resentments about past wrongs, aggressive territorial expansion, or religiously inspired altruism toward the future. If we think this sort of bargaining is inevitable, we would conclude that climate change is a racial-hatred-related or religious or some other kind of problem. Nations might be inspired by patriotic music and bargain harder because of this inspiration. Climate change would then be a musical problem.

It strikes me that this is not what people mean when they say that climate change is an ethics problem. Instead, they mean it in the first sense given above: that there is a set of ethical views that determines what *obligations* nations have to reduce emissions. Simply because some nations might demand what they view as fair treatment does not convert the problem into an ethics problem.

5.3 IS ETHICS THE RIGHT TOOL?

Suppose, notwithstanding the logical flaws in the ethical arguments made so far and notwithstanding the power of self-interest, that understanding the problem of climate change requires us to invoke justice or ethics. Perhaps we need ethics to ensure that we properly value those who are not at the bargaining table, such as the distant future or nonhuman animals. Even so, climate change is not primarily an ethical problem.

Solving the problem of climate change is about wisely managing the use of a limited and vital resource. The problem is difficult because we use the resource whenever we use fossil fuels, and fossil fuel energy is the basis of modern life. Fossil fuels are used to heat and light our homes, provide transportation, and produce goods. Your morning shower and cup of coffee cause climate change. Solving the problem of climate change is most centrally a problem of finding a way to produce energy that does not result in emissions of carbon dioxide. It will require new infrastructure, technology, and engineering.

Beyond engineering, solving the problem of climate change requires coordinating hundreds of nations and thousands, millions, or possibly billions, of actors each with their own self-interest and their own ethical views. It involves understanding how to reduce or eliminate deforestation in regions of the world that are poor and poorly governed. It requires designing tools for monitoring and enforcing agreements.

Philosophy lacks the tools for analyzing these sorts of problems, such as the engineering, the science, the required incentives, the trade effects, and the wide variety of other effects of climate change policy which all occur in a complex global economic and political equilibrium. Because it is ill equipped for the task, trying use philosophy to design climate change policy will, except by sheer happenstance, lead to bad policies. Ethics can help understand our values and help clarify our thinking, but it is not the right tool for the design of policies, particularly policies as complex and important as climate change policies.

5.4 WHAT ROLE FOR ETHICS?

If we take self-interest seriously enough, there may be no room for ethics. If people or nations only pursue their self-interest, nothing ethics can say will influence the outcome. We should just get on with solving the problem: hire the engineers, fix the relevant laws, negotiate treaties, and so forth.

While this view as has a certain attractiveness, it is too strong. Consider the following contributions that ethics can make. First, when I noted that self-interest by itself leads to stringent emissions reductions and can point the way toward wise use of common resources, there was an implicit evaluation. The reason for pointing out these effects of the pursuit of self-interest is that they are desirable. Saying something is desirable requires some sort of moral stance that was not made explicit. The argument that we do not need ethics because self-interest produces desirable results is itself an ethical argument.

Fair enough, and if this is how ethics informs climate policy, I don't object. On the other hand, it is not clear that we always need serious ethical reflection to know when pursing self-interest is a good idea. To a great extent, if we are behaving in a way that is contrary to our own self-interest, we do not use, or need, ethics to tell us what to do. We don't say, "Stop hitting yourself in the head with a hammer. It is unethical." We say, "Stop hitting yourself with a hammer. You will hurt yourself." The appeal is purely to self-interest. If this is an ethical claim, it is not an important or interesting one.

Second (and relatedly), the things we think of as being in our own self-interest might be determined to some extent by ethical or moral concerns. We are not born knowing precisely what we want in life. Some of our goals and views about what makes a good life are informed by philosophy. The extent to which philosophy influences perceived self-interest is a difficult empirical question. Other influences, such as base instinct, family, religion, culture, role models, and so forth might matter significantly more. Patriotic music might be more important in determining how people behave than ethics even if ethics provides an element of our understanding of our needs and goals.

The influences on how nations act will be more complex than for people because nations have to aggregate the views of their citizens. In the case of climate change, nations may perceive it in their own interest to be treated fairly and to treat others fairly.

We cannot easily resolve the extent to which ethics and philosophy affect perceived self-interest. The answer may vary widely by culture and, for nations, by the type of government. Nevertheless, I do not believe that any amount of ethical reflection will convince many nations that the various climate policies that are not feasible are in fact in their self-interest. I do not, for example, believe that the United

States will ever view it as in its self-interest to enter into the sort of treaty that Peter Singer suggests because doing so would require it to transfer trillions of dollars to others nations, all so that it can continue doing what it can do without a treaty.

More importantly, relatively narrow notions are sufficient to motivate action on climate change. Not limiting climate change will directly hurt us, our children, and our grandchildren in straightforward ways. Climate change threatens our food supplies, our cities through storms and sea level rise, and, generally, our lives as we have come to know them. While I believe in a broad notion of self-interest and well-being, we need not engage in debates about exactly what this means to know that we should want to limit climate change.[22]

Third, I relied on philosophical arguments when I made the claim that feasibility is an appropriate limitation on the types of policies we should consider. The argument that feasibility is a limitation on philosophical claims is itself a philosophical claim.

Finally, in prior work with Eric Posner, I endorsed a cosmopolitan view of distributive justice for purposes of analyzing aspects of climate change.[23] The claim is that the world would be a better place if wealth, income, education, opportunity, and other valuable items were more equally distributed (without a significant reduction in the total amount). As a result, people may have obligations in distributive justice to people in other countries. This is a controversial view, subject to significant and wide-ranging debate in the philosophical literature.

An implication of a cosmopolitan view of distributive justice is that the worse off a nation is, the greater the obligation to help. A wealthy nation may only have a modest obligation to help a nation that is just a little bit worse off, but may have a substantial obligation to help a desperately poor country. Climate change threatens to make poor countries even worse off than they are today, in which case wealthy nations would have even stronger obligations to help. Similarly, limiting a poor country's access to cheap fossil fuels in an attempt to minimize global emissions might make it worse off, again triggering greater obligations to help. As I will discuss in chapter 7, however, these obligations should be met in the most effective way possible. A cosmopolitan view of distributive justice tells us our overall set of goals but not how to meet them. It does not require wearing climate change blinders.

In the end, I think it is important that climate policy engage with philosophy. I am not a philosopher. My expertise is in policy. I work mostly with scientists and economists on policy issues such as the design of a carbon tax, the mix of regulatory instruments such as taxes and cap-and-trade systems, and the effects of fossil fuel infrastructure on fuel switching costs. I was asked to write my part of this book to think about the extent to which the work I do in climate policy needs to engage with philosophy or can gain from such an engagement.

I think most of the problems we need to address to solve the problem of climate change are not particularly philosophical. For example, we need to (it is in our self-interest to) invent technology to produce inexpensive clean energy. We need to be able to make that technology available widely and ensure that it is used instead of fossil fuel. Engineers and scientists will think about the details of the technology. Lawyers and economists will think about what sorts of policies will promote the needed technological changes such as changes to the patent rules or subsidies for new technologies. Political scientists might consider the design of treaties and how they can be enforced. And so forth.

I would not have said that these, the central challenges in addressing climate change, are ethical. Although there may be ethical components, I think we need to do these things to save our own necks. Nevertheless, engaging with philosophers can help us understand when values have implicitly entered the analysis and help us clarify thinking about whether these values are the right ones.

5.5 BACKGROUND AND HISTORY

To understand the role that ethics should play in climate change policy, it is helpful to know something of the history of climate negotiations and the role that it has played.

The first and most important global climate change treaty is the Framework Convention on Climate Change, signed in 1992.[24] Virtually all nations, including the United States, are signatories. The Framework Convention is, to a great extent, aspirational. It does not impose specific obligations to reduce emissions. It is a framework for future negotiations not an emissions reduction treaty. Nevertheless, as a framework, it establishes a set of principles that have been important in all subsequent negotiations.

The first principle adopted in the Framework Convention is as follows:

> The Parties should protect the climate system for the benefit of present and future generations of humankind, on the

> basis of equity and in accordance with their common but
> differentiated responsibilities and respective capabilities.
> Accordingly, the developed country Parties should take the
> lead in combating climate change and the adverse effects
> thereof.[25]

The developed country parties, listed in Annex I of the Convention and usually called Annex I countries, include Western Europe, the United States, Canada, Australia, and most of the countries in Eastern Europe and the former Soviet Union. China, India, Brazil, South Korea, and other fast growing nations are not developed country parties.

The Framework Convention gives reasons for this approach. It notes that "the largest share of historical and current global emissions of greenhouse gases has originated in developed countries, that per capita emissions in developing countries are still relatively low and that the share of global emissions in developing countries will grow to meet their social and development needs." The Convention also affirmed that "responses to climate change should be coordinated with social and economic development in an integrated manner with a view to avoiding adverse impacts on the latter, taking into full account the legitimate priority needs of developing countries for the achievement of sustained economic growth and the eradication of poverty."

The Framework Convention is rooted in theories of justice. It requires equity. It gives priority to growth in developing countries reflecting notions of distributive justice.[26] It anticipates differential obligations on countries based on past behavior reflecting notions of corrective justice. It encompasses the three core theories of justice that I will examine here: equity, distributive justice, and corrective justice.

These theories of justice, particularly as embodied in the notion of common but differentiated responsibilities, were central to subsequent negotiations. The first negotiation following the Framework Convention—the so-called Conference of the Parties—was in Berlin in 1995. It led to an agreement known as the Berlin Mandate. The Berlin Mandate took a strong view of the meaning of the first principle of the Framework Convention and the term "common but differentiated responsibilities." Under the Berlin Mandate, Annex I countries were to agree to specific reductions in emissions while other countries would have no obligations whatsoever. China, India, Brazil, and similar countries were to be free to increase their emissions without limit. This approach has been called the dichotomous distinction because it draws a sharp line between wealthy countries that have to reduce emissions and developing countries that do not.[27]

The dichotomous distinction is hard to justify other than on principles of justice or ethics. A treaty focused on finding the most cost-effective way of preventing climate change would not allow emissions to increase in large parts of the world. A treaty focused on the most cost-effective approach to climate change would seek to reduce emissions at the lowest possible cost (or to reduce emissions the most at a given cost). This would mean that if it is easier or less expensive to reduce emissions, or stop their increase, in a poor country, we would do so. A treaty focused on cost-effective solutions to climate change would not decide where to reduce emissions based on the wealth of the polluter or their past behavior. The choices in the Berlin Mandate were instead based on notions of justice. The parties made a conscious choice to sacrifice pure climate change goals for goals based on distributive justice, corrective justice, and equity.

The dichotomous distinction was formally adopted in the 1997 Kyoto Protocol. Under the Kyoto Protocol and subsequent related accords, Annex I countries (relabeled as Annex B under the Protocol but with minor exceptions the same list of countries) agreed to specific obligations to reduce emissions and to timetables for meeting the obligations. Non-Annex I countries were free to continue to increase emissions without restriction. The Kyoto Protocol was, and remains as I write, the only treaty in which nations agreed to binding obligations to reduce emissions.[28] Like the Berlin Mandate, it is hard to understand the Kyoto Protocol without principles of justice. It does not seek to reduce emissions in the most cost-effective way. Instead, it chooses where to reduce emissions based on principles of justice.

The United States Senate responded by passing a resolution unconditionally rejecting this approach. In what was known as the Byrd-Hagel Resolution, the Senate stated:

> It is the sense of the Senate that the United States should not be a signatory to any protocol to, or other agreement . . . which would mandate new commitments to limit or reduce greenhouse gas emissions for the Annex I Parties, unless the protocol or other agreement also mandates new specific scheduled commitments to limit or reduce greenhouse gas emissions for Developing Country Parties within the same compliance period.[29]

The resolution listed a number of reasons for this view. A central one is concern about the interests of the United States, expressed as follows:

> [T]he Senate strongly believes that the proposals under negotiation, because of the disparity of treatment between Annex I Parties and Developing Countries and the level of required emission reductions, could result in serious harm to the United States economy, including significant job loss, trade disadvantages, increased energy and consumer costs, or any combination thereof.

The Byrd-Hagel Resolution passed by a vote of 95-0. Every member of the Senate who voted that day, regardless of their commitment to the environment, their green credentials, or their party, voted for the resolution. The Clinton Administration did not bother to submit the Kyoto Protocol to the Senate for ratification. It had no chance of being ratified.

One reading of the Byrd-Hagel resolution and the US failure to ratify the Kyoto Protocol is that the United States failed to act ethically. The United States was too concerned with its own self-interest. The call for more ethical action on climate change could then be seen as a call for the United States to live up to its ethical obligations. Perhaps the Byrd-Hagel resolution is evidence that we need more ethics, not less.

Additional urging of ethics, however, will not work to get the United States to ratify a treaty that looks anything like the Kyoto Protocol. The United States will not agree to a treaty that imposes obligations to reduce emissions on the United States but not on China, India, and other fast-growing developing nations. A treaty of this sort, without more, is not in the interests of the United States.[30] Moreover, a treaty of this sort is not sufficient to stop climate change, so it fails to fulfill its basic purpose. And, as I will argue in chapter 7, it is not required by principles of justice, which means that the United States need not agree to such a treaty based on ethical concerns. The Berlin Mandate and the Kyoto Protocol are prime examples of the failure of applying ethics to climate change.

Notwithstanding the refusal of the United States to ratify the Kyoto Protocol, the Protocol did eventually receive enough support to go into effect. Its primary effect has been in the European Union, which has enacted policies to reduce emissions. These policies have met with some success. EU emissions by some measures are down modestly since the Protocol took effect, and the European Union is clearly the world's leader in climate change policy.

The initial period for the Kyoto Protocol expired in 2012, but as part of the Framework Convention negotiations in 2011 in Durban, South Africa, signatories agreed to a second five-year commitment period. While the Kyoto Protocol was extended in Durban, the negotiations there also represented a dramatic break with the dichotomous distinction approach of the Kyoto Protocol.

By the time of the Durban negotiations in 2011, it was relatively clear to most that the Kyoto approach had failed. There was no chance that the United States would agree to it. Canada had ratified the treaty but eventually dropped out when it became clear that it would not meet its targets and would be subject to penalties if it stayed in. Japan, New Zealand, and Russia did not agree to take on additional commitments in the second phase of the treaty. Australia seems to flip positions with every change in government.

More centrally, the world had changed dramatically since 1992 when the common but differentiated responsibilities approach was formulated and since 1995 when it was interpreted as establishing a dichotomous distinction. China went from a modest emitter to the world's dominant emitter. Countries such as India have grown rapidly and their emissions correspondingly increased. By 2010, non-Annex I countries accounted for 62% of global emissions and were growing quickly. An approach that did not address these emissions, an approach that allowed them to increase without restrictions, held no hope of stabilizing the carbon dioxide concentrations. Moreover, the levels of CO_2 in the atmosphere were by 2010 much higher than in 1992, leaving less room for emissions increases by anyone before concentrations reached the level at which they would cause serious harms.

The negotiators in Durban, therefore, took a different approach. They agreed that future climate treaties should require the highest possible mitigation efforts by *all* Parties."[31] If a future agreement follows the Durban approach, there will no longer be a sharp distinction between developed and developing countries. While countries may be allowed to take different paths to emissions reductions, no country with more than modest emissions will be exempt from reduction obligations.

As I write, there is a serious question whether nations will follow the Durban approach or fall back on the approach of the Berlin Mandate. Negotiators from developing countries have argued for a return strong differentiation between developed and developing nations. Some have suggested that they would reject any treaty that differs from the Berlin Mandate's interpretation of common but differentiated responsibilities. The United States says it that will not agree to any approach that does not restrict emissions from China and similar developing countries. And even if the United States agreed to such a treaty, if developing countries continue to increase emissions, we will not be able to stabilize the atmosphere. One hopes that this is all just posturing to gain advantage in negotiations and that the nations of the world will agree to do what is in their self-interest, to wisely manage the limited resource.

I close this chapter with some basic data on emissions, data which is central to applying theories of justice to climate change. To get a sense of how the world now compares to the world of the Framework Convention and the Berlin Mandate, we can compare emissions in 1992 to those of today. In 1992, global emissions including land use change were about 33 billion tons of CO_2 or equivalents (such as emissions of methane or nitrous oxide). In 2010, emissions were 45 billion tons. Most of this growth has come from fast developing countries like China (whose emissions have almost tripled) and India (whose emissions have doubled). Table 5.1 shows the top 15 emitters ranked in order of their 2010 emissions.

Where are we headed if we do not adopt policies to control emissions? Prediction is hard, especially when it is about the future. Nevertheless, in the absence of policies designed to reduce emissions, almost all of the emissions increases over the next twenty-five years are expected to come from developing countries. Emissions from developed countries are expected to be flat or to decline modestly.

With this background, we are ready to turn to climate policy and where (morally corrupt?) self-interest leads us.

Table 5.1

EMISSIONS OF GHG'S INCLUDING LAND USE CHANGE, VARIOUS YEARS

(DATA FROM WRI CAIT)

	Country	1992	2010	Percent increase	Cumulative 1990–2010
1	China	3,362	9,387	179%	112,668
2	United States	5,730	6,254	9%	130,616
3	European Union (28)	4,933	4,386	–11%	99,706
4	India	1,120	2,304	106%	31,738
5	Russian Federation	3,035	2,134	–30%	50,136
6	Indonesia	1,134	2,033	79%	32,183
7	Brazil	1,757	1,393	–21%	39,070
8	Japan	1,146	1,120	–2%	24,991
9	Canada	621	842	35%	16,361
10	Germany	1,037	827	–20%	20,011
11	Mexico	469	706	50%	12,100
12	Iran	310	695	125%	9,871
13	South Korea	316	630	99%	9,657
14	Australia	437	592	36%	10,987
15	United Kingdom	728	579	–20%	13,880
	Rest of the World	12,022	14,877	24%	273,232

Notes

1. I thank Thomas Christiano, Martha Nussbaum, Mike Seidman, and workshop participants at Ohio State University, The University of Arizona, The University of Chicago, and Florida State University for comments and suggestions.
2. Unless otherwise stated, all of the emissions data used here are from http://cait.wri.org/ (accessed Oct 3, 2014). The estimates use a standard conversion factor to include greenhouse gases other than CO_2.
3. Stephen Gardiner, "Climate Justice," in *The Oxford Handbook of Climate Change and Society*, eds. John S. Dryzek, Richard B. Norgaard, and David Schlosberg (Oxford: Oxford University Press, 2011), 309.
4. Stephen M. Gardiner, *A Perfect Moral Storm: The Ethical Tragedy of Climate Change* (New York: Oxford University Press, 2011), 68.
5. Peter Singer, *One World: The Ethics of Globalization* (New Haven: Yale University Press, 2002).
6. den Elzen, Michel, Jan Fuglestvedt, Niklas Höhne, et al. "Analysing Countries' Contribution to Climate Change: Scientific and Policy-Related Choices," in "Mitigation and Adaptation Strategies for Climate Change," eds., special issue, *Environmental Science & Policy* 8, no. 6 (2005): 614–636; Michel G. J. den Elzen, et al., "Countries' Contributions to Climate Change: Effect of Accounting for All Greenhouse Gases, Recent Trends, Basic Needs and Technological Progress," *Climatic Change* 121, no. 2 (2013): 397–412; Michel den Elzen and Michiel Schaeffer, "Responsibility for Past and Future Global Warming: Uncertainties in Attributing Anthropogenic Climate Change," *Climatic Change* 54, no. 1–2 (July 1, 2002): 29–73.
7. Henry Shue, "Subsistence Emissions and Luxury Emissions," *Law & Policy* 15, no. 1 (1993): 537. Paul G. Harris, *World Ethics and Climate Change: From International to Global Justice* (Edinburgh: Edinburgh University Press, 2010), 131. James Garvey, *The Ethics of Climate Change* (London: Continuum, 2008), 81.

8. The arguments I make here are related to, although not necessarily the same as, the arguments I made in prior work, most importantly Eric Posner and David Weisbach, *Climate Change Justice* (Princeton, NJ: Princeton University Press, 2010). Other related work includes Eric A. Posner and Cass R. Sunstein, "Climate Change Justice," *Georgetown Law Journal* 96, no. 5 (2008): 1565–1612, and Eric A. Posner and Cass R. Sunstein, "Should Greenhouse Gas Permits Be Allocated on a per Capita Basis?," *California Law Review* 97, no. 1 (2009): 51–93. There are a number of differences between the approach taken here and that taken in the Posner and Sunstein papers and to the extent there are differences, their views should not necessarily be taken to be mine.

9. Most philosophers supported this approach at the time. As Gardiner stated in 2004, "[T]here is a surprising convergence of philosophical writers on the subject: they are virtually unanimous in their conclusion that the developed countries should take the lead in bearing the costs of climate change, while the less developed countries should be allowed to increase emissions for the foreseeable future." See Stephen M. Gardiner, "Ethics and Global Climate Change," *Ethics* 114, no. 3 (2004): 579. Gardiner himself does not support this approach; in Gardiner, "Climate Justice," 314, he wrote "[i]t is now clear that the developing nations will have to constrain their own emission quickly and significantly . . . From the point of view of justice, the consensus appears to be that this only increases the burden on developing [*sic*] countries to assist in other ways.")

10. The discussion in the next few paragraphs paraphrases the discussion in Eric A. Posner and David Weisbach, "International Paretianism: A Defense," *Chicago Journal of International Law* 13, no. 2 (2013): 347.

11. John Rawls, *Political Liberalism* (Columbia University Press, 1993), 450–457.

12. Thomas Nagel, "Moral Conflict and Political Legitimacy," *Philosophy & Public Affairs* 16, no. 3 (1987): 215–240, 218.

13. Elizabeth Anderson, *The Imperative of Integration* (Princeton, NJ: Princeton University Press, 2010).

14. See Jack L. Goldsmith and Eric Posner, *The Limits of International Law* (Oxford: Oxford University Press, 2005); Eric A. Posner and Alan Sykes, *The Perils of Global Legalism* (Chicago: University of Chicago Press, 2009); Posner and Weisbach, "International Paretianism."

15. Cass R. Sunstein, "Of Montreal and Kyoto: A Tale of Two Protocols," *Harvard Environmental Law Review* 31, no. 1 (2007): 1.

16. Beth A. Simmons, *Mobilizing for Human Rights: International Law in Domestic Politics* (Cambridge University Press, 2009).

17. Gardiner, *A Perfect Moral Storm*, 68.

18. Climate change will also harm people in the more distant future. One of the central issues at the intersection of ethics and economics is how to value the harms to people living in the distant future. The literature on this topic, usually known as discounting, is vast. Due to space limitations I will not discuss discounting here. My views (and the reasons why discounting is not centrally an ethical issue) can be found in David A. Weisbach and Cass Sunstein, "Climate Change and Discounting the Future: A Guide for the Perplexed," *Yale Law and Policy Review* 27, no. 2 (2009): 433–457; Posner and Weisbach, *Climate Change Justice* (Princeton, NJ: Princeton University Press),144–168; and Elisabeth J. Moyer, et al., "Climate Impacts on Economic Growth as Drivers of Uncertainty in the Social Cost of Carbon," *Journal of Legal Studies* 43, no. 2 (2014): 418–421.

19. The extent to which common resources are protected (or, equivalently, public goods provided) without government is unclear. For many years it was thought that the government was needed to provide lighthouses because a lighthouse owner cannot charge passing ships for its use. Ronald Coase, however, showed that there was an extensive network of private lighthouses. See R. H. Coase, "The Lighthouse in Economics," *Journal of Law & Economics* 17, no. 2 (1974): 357. Elinor Ostrom won the Nobel Prize in Economics for showing that cooperation to govern the commons is far more extensive than previously thought; see Elinor Ostrom, *Governing*

the Commons: The Evolution of Institutions for Collective Action (Cambridge, UK: Cambridge University Press, 1990).

20. Jared Diamond, *Collapse: How Societies Choose to Fail or Succeed* (New York: Viking Penguin, 2005).

21. Henry Shue, *Climate Justice: Vulnerability and Protection* (Oxford: Oxford University Press, 2014), 4. Gardiner, *A Perfect Moral Storm*, 419.

22. The argument in the text focuses on people alive today and their immediate descendants. The papers cited in note 16 discuss why self-interested choices fully take future generations into account.

23. Posner and Weisbach, *Climate Change Justice.*

24. *United Nations Framework Convention on Climate Change,* June 3–14, 1992.

25. *Framework Convention* (1992), Article 3, paragraph 1.

26. This is also seen in the second principle which states "the specific needs and special circumstances of developing country Parties, especially those that are particularly vulnerable to the adverse effects of climate change, and those Parties, especially developing country Parties, that would have to bear a disproportionate or abnormal burden under the Convention, shall be given full consideration."

27. Joseph E. Aldy and Robert N. Stavins, "Climate Negotiators Create an Opportunity for Scholars," *Science* 337, no. 6098 (2012): 1043–44.

28. The Montreal Protocol, which limits emissions of ozone depleting substances, has actually had a much larger effect on greenhouse gas emissions than the Kyoto Protocol because many of the banned ozone-depleting substances were strong greenhouse gases. See Guus J. M. Velders, et al., "The Importance of the Montreal Protocol in Protecting Climate," *Proceedings of the National Academy of Sciences* 104, no. 12 (2007): 4814–4819.

29. Byrd-Hagel Resolution, S. Res. 98, 105th Congress (1997).

30. One possibility is for the United States to agree to move first with the understanding, either implicit or explicit, that developing countries will soon follow. A treaty of this sort might be in the interest of the United States. My own

view is that I wish the United States had ratified the Kyoto Protocol even though I think the treaty design, shaped as it was by ethical concerns, was terribly flawed. Starting fuel switching sooner rather than later is, I believe, cost saving, and ratifying the Kyoto Protocol might have accelerated this process in the United States. If we eventually have to move to renewable energy, we might as well start now and do it slowly rather than wait and try to do it all at once. Moreover, it might have made negotiations with developing countries over their commitments easier (although this is highly speculative). By accelerating fuel switching and helping to induce developing countries to agree to emissions reductions, ratifying the Kyoto Protocol might well have been in the United States's self-interest.

31. United Nations, UNFCCC, "Durban Platform for Enhanced Action 1/CP.17" (2011).

6

Climate Policy and Self-Interest

TO UNDERSTAND HOW justice and ethics apply to climate change, it is useful to consider the major alternative: self-interest. What should we do about climate change if we act purely to save our own necks?

I will seek to show here that, under almost any plausible assumptions about climate change, it is in our self-interest to start reducing emissions now, on a global basis, and to reduce emissions to near zero in the not-too-distant future. This conclusion is robust to a broad range of assumptions about the science and the economics of climate change. It follows solely from the simple self-interest of people who are alive today, their children, and their grandchildren. It holds, and is possibly strengthened, when we consider the uncertainty we face about the harms from climate change and the costs of avoiding these harms.

There are five main points that together establish this conclusion:

- Net emissions must eventually be reduced to zero or near zero because temperatures keep on increasing as long as emissions are positive, and because there is a limit to tolerable temperature increases.

- The time limit is not too far off. While there is great uncertainty about the appropriate limit on temperature increases, most mid-range targets will be reached by the middle or end of this century, during the lives of people alive today, their children, and their grandchildren. If emissions are not reduced to near zero, these individuals, people living today and their immediate relatives, will suffer. Climate change might have primarily been a long-term problem when negotiations began in the early 1990's, but because emissions have increased rapidly since that time, it is now also a medium-term problem.
- Near-zero emissions in the not-too-distant future means we have to start reducing emissions now. The reason is that reducing emissions to zero or near zero means replacing the global energy system, which is currently based on fossil fuels. Replacing the global energy system is an enormous task. The system is vast, probably the largest and most complex system humans have ever built. We have been building it for more than one hundred years. The only feasible way to replace it is to start now or in the near future.
- Reductions have to be global. Zero emissions means zero. Nobody will be able to emit carbon dioxide, including people in developing countries. Developing countries are installing a massive new fossil fuel infrastructure. If this installation continues, it will make achieving zero emissions in the relevant time frame difficult or impossible. Both developed and developing countries have to start reducing emissions now or in the near future.

- Uncertainty about the effects of climate change strengthens these conclusions because the uncertainty is not symmetric: if we do nothing or act too slowly, the bad cases if things turn out worse than expected are far worse than the good cases are good if things turn out better than expected.

These conclusions sound stark, almost extreme, when compared to the glacial pace of climate change negotiations. They are not, however, based on unsupported apocalyptic visions of extreme environmentalism. They are, unfortunately, based on conservative assumptions.

6.1 EMISSIONS MUST GO TO ZERO

We can think of greenhouse gases—carbon dioxide (CO_2) and a number of other gases such as methane and nitrous oxide—as acting like a blanket. They cover the Earth and keep the warmth in, heating up the surface where we live. The greater the concentration of greenhouse gases in the atmosphere, the thicker the blanket, and the warmer we are.

Until the Industrial Revolution, humans had little impact on the climate. Average surface temperatures varied somewhat over the centuries due to natural variation, but within the last ten thousand years—since the beginning of agricultural civilization—they have been a relatively constant 15°C.

The Industrial Revolution threatens to change that. The Industrial Revolution was based on fossil fuels.

Technologies such as the steam engine were able to convert the energy stored in fossil fuels into motion, enabling the mechanization of many tasks. The beginning of the industrial revolution was powered by coal. Over time, economies have diversified their fuel mix and now rely heavily on three fossil fuels: coal, oil, and natural gas. All three, when burned to produce energy, release CO_2 into the atmosphere.

As a result of industrialization and related changes such as population increases, global emissions went from near zero in 1850 to around 34 billion tons of greenhouse gases in 1992 and 45 billion tons in 2010. Globally, about 65% of this total is from fossil fuels, 12% from deforestation, and the rest from emissions of other greenhouse gases such as methane and nitrous oxide, largely from agriculture. In developed countries such as the United States about 80% of emissions come from fossil fuels because these countries have largely eliminated deforestation.

The resulting increase in the thickness of the greenhouse gas blanket will increase global temperatures. So far, temperatures have increased by about 1°C from their preindustrial level. If we continue to emit, temperatures may go up anywhere from around 2°C to as much as 6°C or even more, depending on how much we emit and how sensitive the atmosphere turns out to be to greenhouse gases. These are global averages. Many places, such as northern land masses, may heat up far more, and other places, such as the area over the ocean, may heat up less.

There are three key features of the greenhouse gas blanket that lead to the conclusion that we must eventually reduce emissions to zero. The first is that CO_2 in the atmosphere is effectively permanent. The CO_2 we emit today will continue to influence the climate for tens or even hundreds of thousands of years.[1] There is, moreover, no way to remove it, at least using any technology that we have now or that is foreseeable.

The second feature is that temperatures continue to go up when we emit more. As the greenhouse gas blanket gets thicker, we get warmer.

The amount of warming we will get is uncertain, and the science enormously complex. Notwithstanding years of work, we cannot pin it down. There is, however, a simple way to understand the core relationship between emissions and temperatures: temperatures go up linearly with the total amount of CO_2 emitted in the past.[2] All we need to know is the cumulative emissions of carbon to know what the likely temperature increase will be.

Figure 6.1 is a simplified representation of this relationship. The x-axis shows cumulative emissions of carbon, the sum of all emissions in the past, regardless of when they occurred.[3] The y-axis shows the expected increase in global average temperatures. The central black line reflects the current best estimate of how sensitive temperatures are to emissions, a value known as climate sensitivity. Using this estimate, there is a 50/50 chance of the specified temperature increase for a given level of cumulative emissions. For example, for cumulative emissions of 1 trillion tons of carbon, we have a 50/50 chance of a 2°C temperature increase. For a cumulative emission of 1.5 trillion tons, there is a 50/ 50 chance of a 3°C temperature increase.

We do not know how much temperatures will increase for a given level of emissions, a value known as the climate sensitivity. The light gray lines reflect uncertainty regarding climate sensitivity. If we get lucky and the climate is relatively insensitive (the bottom gray line), we can emit

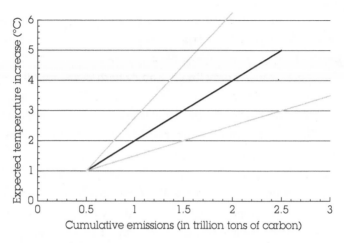

FIGURE 6.1 Temperature as a Function of Cumulative Emissions.

1.5 trillion tons of carbon before we have a 50/50 chance of a 2°C temperature increase. If we are unlucky and the upper-gray line represents the true climate sensitivity, we can emit only about 800 billion tons before temperatures increase by 2°C.

Figure 6.1 points to a key conclusion. Whatever limit we set on temperature increases, emissions have to stop once we meet this target. For example, if the target is 2°C, we can only emit 1 trillion tons of carbon. Any more than that would lead to a likely temperature increase greater than 2°C. The same holds for any other limit we set. A 4°C limit means we can emit at most 2 trillion tons.

Said another way, on human timescales, the atmosphere is a nonrenewable resource. For a given temperature increase, it can hold a fixed amount of carbon and no more. The atmosphere is not like agricultural land, which can be replenished, or fisheries or forests, which if left alone, regrow. It is a strictly limited resource. As a result, unless we decide to let temperatures increase indefinitely, emissions have to go zero. This is true even given uncertainty about the climate sensitivity. Regardless of which line we are on— the upper or lower gray line or somewhere in the middle— emissions have to go to zero to stop temperature increases.[4]

The third central feature of climate change is that there is a limit to tolerable temperature increases. The harms from climate change are even more uncertain than the extent of temperature increases, but they will go up rapidly as temperatures increase. We have already experienced about a 1°C temperature increase. There have been some harms but they have not been extreme. Economic growth has continued. Problems other than climate change dominate our attention. At the other end of the spectrum, temperature increases of 5°C or 6°C would almost certainly be catastrophic. The last time temperatures were 6°C warmer was around 40 million years ago. There was no ice anywhere on Earth and sea levels were an astounding 60 meters higher than they are today.[5] Many ecosystems that depend on the current climate would collapse. To get a sense of the magnitude, the last ice age, when much of North America and Europe were covered with mile high glaciers, was only 6°C colder than today. There is no possibility that policies that would lead to this sort of warming are desirable. There is a limit to tolerable temperature increases.

These facts—that if we continue to emit greenhouse gases, temperatures keep on going up, that the harms will likely go up rapidly as temperatures increase, and that there is a limit to tolerable temperature increases—mean that we eventually have to reduce emissions to zero. This is true even if the temperature turns out to be relatively insensitive to greenhouse gases and even if the harms turn out to be on the lower end of the possibilities. The only thing that varies with these factors is the total allowable emissions before we must stop. When we think about the ethics of climate change we have to think of it as the ethics of using a vital, nonrenewable resource.[6]

6.2 SOON

Reducing emissions to zero might not be that big a problem if we had hundreds of years to do so. The problem is that we must reduce emissions to zero in the near future.

We cannot put a precise date on when emissions need to be zero. There is considerable uncertainty about the extent of temperature increases, harms from those increases, and the costs of reductions. No human has ever lived in a world where temperatures are on average, say, 3°C or 4°C warmer than they are today, not to speak of 5°C or 6°C, so we have little ability to predict what such a world would be like. We also do not know the costs of reducing emissions.

Current global agreements call for limiting temperature increases to 2°C. This may be unrealistic and perhaps a more realistic goal is 2.5°C or 3°C. Targets in this range, however, are not very far away. Consider Figure 6.1. Suppose our target is 2°C, we use the central estimate for the climate sensitivity, and we accept policies that give us only a 50% chance of meeting this target. This means that we can emit at most one trillion tons of carbon. We have already emitted about 580 billion tons of carbon and are emitting about 9 billion tons more each year (which means more than 30 billion tons of carbon dioxide). Even if the pace of emissions does not increase (and it has been increasing rapidly), we would hit the trillion-ton limit sometime between 2050 and 2060. If emissions rates increase, as they are likely to do without a change in policies, we will hit it even sooner with some estimates showing that we hit the trillion-ton limit before 2040.[7]

This timescale is based on accepting a 50/50 chance of exceeding a 2°C target and the central climate sensitivity. Let us see how much time we can buy by making the most optimistic assumptions. Assume that the climate is extremely insensitive to emissions, and say we only get 1°C of warming for a trillion tons of carbon. I am not aware of anyone who believes that the right target is greater than 4°C. It seems likely that the harm from temperature increases will start to go up rapidly at some point and almost certainly by the time we get to 4°C. At this level, we risk the collapse of major agricultural systems, leading to widespread disaster.[8] Let us assume that we are, gulp, willing to accept a 4°C limit. The implied time limit with these assumptions will give us the longest possible period before emissions need to stop.

Even with these assumptions, if we continue on our current path, we hit the target before the end of the century. If we slow emissions, we buy more time because we use up the total more slowly. We can imagine a date sometime in the middle of the next century, and we only get this by assuming the lowest reasonable temperature increase for a given level of emissions, accepting the possibly terrible harms from a 4°C temperature increase, and by slowing emissions soon so that we get to our hard limit more slowly.

We cannot, however, count on these optimistic assumptions. If we end up with a high climate sensitivity (say, 3°C for one trillion tons), limit temperature increases to a 2°C, and do not slow emissions, we hit the limit in 2020, which is effectively tomorrow (or possibly yesterday, depending on when you are reading this).

The right target date is somewhere in this range. While we do not know the date, it is not easy to come up with a calculation that extends the time until we hit the target into the distant future. We need to be thinking about a 100-year horizon, and then only if we start slowing emissions soon.

Climate change is often portrayed as a very long-term problem, but our children and grandchildren will be alive near the end of the century and into the next, the time when we will face 2°C, 3°C, or even 4°C temperature increases. If we use middle-of-this-century targets, many of the current adults will be alive. While climate change will continue to affect people centuries in the future, it is a surprisingly near-term problem. Climate change is about people alive today, their children, and their grandchildren as well as the distant future.

6.3 WE NEED TO START REDUCING EMISSIONS NOW

We have to reduce emissions dramatically, to near zero, by sometime around the end of this century. When do we have to start? Does it make sense to wait or should we start now?

6.3.1 Climate Change Is an Energy Problem

To try to get a handle on how to transition to zero emissions, we need to understand where emissions come from and what it means to dramatically reduce emissions. I will focus on emissions of CO_2. Reducing emissions of other greenhouse gases like methane and nitrous oxide will also be important but the dominant long-lived greenhouse gas is CO_2.

The overwhelming source of CO_2 emissions is fossil fuels. Fossil fuels are chains of carbon molecules (plus other things) stored underground. When we burn fossil fuels to create energy, we take the carbon that had been underground and put it in the atmosphere.

Solving the problem of climate change means not doing this, not taking carbon from underground and putting it in the atmosphere. It might be possible to prevent the carbon from fossil fuel use from entering the atmosphere by capturing it when we burn the fossil fuels and then storing it underground. So far this technology has proven expensive and implementing it at scale appears to face possibly insurmountable problems because of the difficulties of transporting the CO_2 and of finding safe places to store it. Absent feasible capture technology, solving the problem of climate change means eliminating the use of fossil fuels.

Unfortunately, eliminating or even substantially reducing the use of fossil fuels is going to be difficult. The reason is that energy is central to the global economy and fossil fuels are the central source of energy. Energy's sheer pervasiveness and reliability makes it easy to ignore, but almost everything we do relies on energy. We take it for granted that our homes are heated, cooled, and lit, and we can get to work, take hot showers, refrigerate our food, have concrete and steel to use for construction, and can obtain products from far away. All of these activities rely on energy. It is not too far from the truth to say that the Industrial Revolution and the basis of modern living arose from new ways to transform energy into useful products. It is easy to miss this because our energy system is so utterly reliable and pervasive that it is invisible.

Figure 6.2 shows the connection between wealth and energy. The horizontal axis shows per capita income for 167 countries (on a logarithmic scale, so that equal increments represent equal percent increases in income). The vertical axis shows energy use per person, using a standard unit known as oil equivalents, also on a logarithmic scale.

The graph shows what we might call the iron law of wealth: increased wealth means increased energy use. While there is some dispersion at the low end (we can be poor with different amounts of energy use), and some rich countries manage to be particularly inefficient, nobody escapes the ironclad relationship between energy and wealth. No nation, regardless of its political system, culture, or fantastic environmental values, has discovered a way to be wealthy without energy use.[9] There is nobody in the bottom right hand corner.

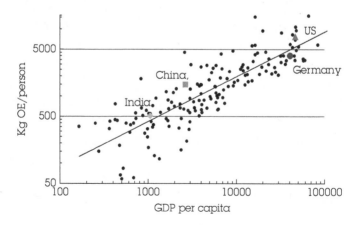

FIGURE 6.2 Income vs. Energy Use (Log Scales) 2007. *Source*: World Bank data.

The bad news is that almost all of this energy comes from fossil fuels. Globally, 87% of energy comes from fossil fuels. Nuclear energy is about 5% and hydroelectric energy is 6.4%. Only 1.6% of energy comes from renewable sources, such as wind or solar.[10]

The central dilemma of climate change is straightforward: emissions have to go to zero while energy use remains high. We have to find a way to replace 87% of the global energy supply with clean energy. This is not just a developed world problem. If developing countries are to have a standard of living the same as the developed world, something we should hope for, everyone will need carbon-free sources of energy. The problem is replacing developed world energy systems and building developing world systems with clean energy.

6.3.2 Energy Transitions Are Slow

This problem—replacing 87% of the existing global energy supply with clean energy and building the developing world's energy system at the same time—involves a massive change in infrastructure. In the United States alone, there is around $6 trillion of long-lived fossil fuel infrastructure which will have to be replaced, (not counting short-lived assets such as 250 million vehicles and millions of commercial and residential furnaces). Power generation and extraction facilities are by far the largest components of this infrastructure. In 2008 in the United States alone there were 6,413 power plants generating 1,075 gigawatts of power, 525,000 crude oil wells, 51,000 miles of crude oil pipelines, and 116,000 miles of refined product pipelines. There were 478,562 gas wells, 20,215 gas gathers, 500 gas processors, 319,208 miles of gas pipeline, and 1.2 million miles of LNG distribution pipelines. All of this has to be replaced. And the United States, although large, is just one country. The global figures are likely to be four or five times as large.

Replacing all of this with renewable energy is not going to happen fast. Even a 100-year horizon—about the longest timeframe we calculated above—is not that long in terms of the needed transformation. To get a sense of how long the process of replacing the fossil fuel infrastructure will take, we can look to the history of energy transitions.

The two major energy technology transitions so far have been the transition from traditional biomass to coal and steam and the transition from coal and steam to oil, gas, and electricity.[11] The process of transition was slow. It took about 130 years from the first use of coal until it became the dominant source of energy, and it took about 80 years for other energy sources—petroleum and gas—to displace coal.

The pace of transition, if anything, might be slower in the future than it was in the past. The system is more built up, there are more people that will be displaced, and people are more dependent on energy so they are less tolerant of disruptions. And we will be replacing the current fossil fuel system with something that may not work quite as well (we would not do it but for the environmental benefits). Past transitions replaced old fuels with better fuels. The history of energy transitions may not tell us what the future will bring, but there is little reason given this history to be optimistic that we can make the shift quickly.

Unfortunately, many strong advocates of action on climate change fail to recognize the problem and instead seem to view the needed transformation as relatively easy if only we had the willpower. For example, Al Gore, in a 2008 address on climate change stated as follows:

> Today I challenge our nation to commit to producing 100% of our electricity from renewal energy and truly clean carbon-free sources within 10 years.

This is sheer fantasy. Actually it is worse than fantasy because it perpetuates the myth that the necessary energy transition can be accomplished in a short time period with just a little gumption. It gives hope to those who want to wait. If we can switch to clean energy in just 10 years, there is less need to act now given that we may have as long as 100 years to make the transformation. The vice president claimed that he was purposefully being ambitious, expanding the set of possibilities, and challenging us. Perhaps a bit of realism is in order.

6.3.3 We Have To Start Now

The choice of when to start this massive transformation involves a trade-off. Much of our energy infrastructure is durable, often lasting 50 to 100 years. If we continue to install fossil fuel facilities, we are likely to have to retire some facilities early, effectively throwing them away. But clean energy is has been getting cheaper over time. If we wait, we might be able to install cheaper clean energy in the future, savings costs.

We cannot be sure of the best timing for transition because we do not know what types of energy technology will be invented in the future. Instead, we have to think about what sorts of bets we want to make.

On the one hand, we can delay switching to clean energy in the hopes that the costs will go down in the future. We can think of this bet as going full steam ahead and then slamming on the brakes at the last minute, replacing the then existing infrastructure with a hoped-for cheap source of clean energy. Everyone will get their own Mr. Fusion generator that runs on garbage.

The cost of losing this bet is that we will have installed massive new fossil fuel facilities that have to be thrown away. Calculations by the International Energy Agency show that these costs will be high.

The IEA estimated that 80% of the cumulative CO_2 emissions between 2009 and 2035 will come from existing capital stock.[12] We have power stations, buildings, factories, refineries, vehicles, furnaces, and the like that exist now or are under construction. These facilities will emit almost all of the allowable emissions under the 1 trillion ton cap. That is, if we simply replace this infrastructure as it wears out with clean energy sources, we still will have used up almost all of the global carbon budget.

Delay allows more infrastructure to be built, locking in more emissions. The same estimate (written in 2011) showed that if we delay emissions reductions until just 2015, around 45% of the global fossil fuel capacity would have to be retired early or refurbished. For every $1 of investment in the power sector that we do not make before 2020, an additional $4.30 would need to be spent later to compensate for delay. If we follow the strategy of delay, the only way to avoid these sorts of costs is to find a dramatically cheaper source of clean energy.

On the other hand, we can start now and go slowly. Starting now means replacing the existing fossil fuel infrastructure with clean energy gradually as the existing infrastructure wears out. And it means meeting new demand with clean energy rather than installing long-lived fossil fuel infrastructure that have to be scrapped.

Gradual replacement not only means we avoid the costs of scrapping perfectly good power plants, refineries, tankers, and other parts of the fossil fuel infrastructure. It also extends the time until we have to stop emitting CO_2 as compared to a full-steam ahead strategy. Reducing emissions now means that we use up less of the total cap, saving more for later.

And starting reductions now actually increases the chance that we have cheap clean energy in the future. The reason is that most of the cost declines in clean energy come in the field, from engineering improvements, not from magical new technologies developed in the lab, like the forthcoming Mr. Fusion. Efficiency improvements come from practice.

We don't know the precise cost-minimizing timing of reductions, but we can think about what sorts of bets we want to make. Starting now and going slowly is the best bet. Perhaps there is a dramatically cheaper source of clean energy just over the horizon, but perhaps not. Unless we are confident that there is, we need to start transitioning now.

6.4 EVERYONE MUST START NOW

To review the argument so far, temperatures keep on increasing as long as we keep on emitting carbon dioxide. Harms from temperature increases will start to go up quickly once we get above a modest level. These two facts combine to mean that we eventually have to reduce emissions to zero. Given the level of past emissions and the current pace of emissions, we are on a schedule to hit plausible temperature targets around the middle or perhaps the end of this century. Under optimistic assumptions, this deadline might be extended into the early part of the next century. After that, emissions will need to be at or near zero. And finally, if emissions are to go to zero on this timescale, we are going to want to start now because of the size of the needed transition in our energy infrastructure.

All of these arguments apply globally and, therefore, to developing as well as developed countries. Zero emissions mean zero *global* emissions, so all countries, developed and developing countries alike, will have to stop emitting CO_2 in the not-too-distant future.

The approach taken in the Berlin Mandate and the Kyoto Protocol would allow developing countries to increase emissions without limit, a conclusion clearly at odds with the necessary emissions reductions. As an alternative, maybe we could allow developing countries to increase emissions in the medium run, say for the next twenty-five years, before they too have to reduce their emissions, eventually to zero. A delay of this sort would allow developing countries to achieve a reasonable level of wealth and allow them the same sort of access to the atmosphere the developed countries had as they grew. This version, while better, is unrealistic as well.

The key reason is that fossil fuel infrastructure is durable. Installing new fossil fuel infrastructure effectively commits a country to the emissions from that infrastructure for its lifetime, which can easily be fifty or even one hundred years. Increasing the size of the fossil fuel infrastructure anywhere makes it more difficult to reach reasonable climate goals. If the increase in fossil infrastructure is large enough—and the planned increases in China and India seem to be—they may make it impossible to reach reasonable climate goals. It is, of course, possible to build a new coal-fired power plant and then shut it down after ten years even though it has forty or fifty years of remaining use, but it is extremely unlikely this would happen. How do leaders of a country explain that they are shutting down a plant that was just built at a considerable cost, works perfectly well, and is providing inexpensive and reliable energy to people who need it? They can't. Once it is built, it will be used. New energy infrastructure instead needs to be clean.

To be sure, there is going to be some new fossil fuel infrastructure, particularly in developing nations. If we—people or nations concerned about climate change—tell India, China, or other fast-developing nations to scrap plans for new fossil fuel energy which they need for their economies to grow, we will simply be ignored. Feasibility concerns run both ways. The more such infrastructure is installed, however, the more difficult it will be to reduce emissions and to hold temperature increases to modest levels.

To get a sense of the need to control emissions in developing countries, we can focus on coal, which is the most important fuel for climate change policy because it is the dirtiest and the most abundant. We can make similar estimates for oil and gas.

Emissions from burning coal are the single large source of emissions, making up 43% of emissions from energy. It is the most polluting of the three fossil fuels, with about twice the emissions per unit of energy of gas. It is used primarily to generate electricity. It is massively abundant and generally easy to mine. There is enough coal at current rates of use to last for well over one hundred years. Reserves of coal potentially hold up to 3.5 trillion tons of carbon.[13] If these reserves are all burned, the implied temperature increase (using a climate sensitivity of 3°C) is 7°C above than today's temperatures. Today's temperatures already represent almost a 1°C temperature increase so burning all of the known coal would produce an 8°C temperature increase. Even if only half of those reserves are recovered and burned, coal alone would produce a temperature increase of 4.5°C above preindustrial levels. If only a quarter of the total coal were burned and 75% of it were left in the ground, we would still greatly exceed the 2°C limit (prior temperature increases of just about 1°C plus 1.75°C more from new emissions from coal). Absent some sort of capture technology, climate stabilization requires that most of the world's remaining coal remain in the ground.

Coal use is declining rapidly in the United States. The problem of new coal infrastructure coal comes largely from China and to a lesser extent India. The growth of Chinese emissions is simply staggering, and the projections even more so. The problem with understanding Chinese emissions is finding the right metaphor to get a sense of the scale.[14] None suffice.

Chinese emissions have grown with their economy, more than tripling since 1990. More than 80% of China's emissions come from coal. China's emissions from coal alone are greater than the total emissions in the United States, the emissions from entire set of OECD countries in Europe, or those of any other nation in the world. China's share of coal production is roughly four times Saudi Arabia's share of oil, yet it now imports coal from Indonesia, Australia, and Vietnam because its domestic production cannot meet demand.

Dieter Helm, a climate change scholar at Oxford, estimated the proposed additions to China's coal infrastructure. He estimates that under the current expansion plans for this decade China will be adding two additional large coal power plants *per week*.[15] Each year China adds new capacity equal to the entire installed electricity capacity of the British electricity system. China's energy is far more reliant in coal than Britain's, so this additional capacity will produce more than double all of Britain's emissions each year.

The World Resources Institute did an inventory of proposed coal-fired power plants (i.e., plants where construction has not yet begun but permits have been applied for).[16] They identified 1,199 proposed plants with a total installed capacity of 1,400 gigawatts. Of these, 560 gigawatts are in China, or 40% of the global total.

Helm produced similar estimates for India. About 70% of India's electricity is from coal. About one-quarter of India's population is without regular access to electricity and in part to supply these people with power, and India plans a 30% increase in its use of coal by 2016. This amounts to about one new power plant each week.

Combined, China and India could be adding three new large coal-fired power plants each week for the coming decade. This makes up 77% of the proposed new coal power capacity for the entire world. The implications for climate change cannot be overstated. These power plants will have useful lives of fifty years or more and would essentially lock in emissions for that time period. It is not realistic to believe both that developing countries can install this sort of fossil fuel infrastructure over the coming decades and that we can keep global temperature increases to modest levels. A claim that developing nations have a right to install this sort of coal infrastructure is an admission that climate change should not be contained because both cannot happen. I don't know where such a right would be said to come from but it is at odds with realistic climate goals.

6.5 UNCERTAINTY STRENGTHENS THESE CONCLUSIONS

The conclusions above were stated without repeated hedges about the vast uncertainty surrounding the effects of climate change. Climate science and climate change policy, however, are plagued by uncertainty. We do not know how much temperatures will go up with emissions. The range is large. We know even less about the physical effects of temperature increases. Major issues such as the extent and speed of the melting of Greenland's and the Antarctic ice sheets defy understanding. We also do not know the economic and social harms that these effects will cause. And we do not know the costs of reducing emissions. It is not wrong to say that the core problem of climate change is one of making choices in the face of deep, irreducible uncertainty.

To develop firm conclusions in the face of uncertainty, I tried to consider the best and the worse cases given what we know and see what conclusions follow. If, for example, even under the best case (i.e., low climate sensitivity) temperatures continue to increase with emissions and that at some point the resulting harms will be intolerable, we know that emissions must eventually be reduced to zero.

It is important, however, to understand whether and how our uncertainty about the effects of climate change might affect these conclusions. Perhaps if we really do not know what is going to happen, we should be cautious about spending the substantial resources needed to transform our currently well-functioning energy system. We have plenty of other problems to solve in the meantime. One hundred years is a long time. Maybe we should stay the course and hope for inexpensive clean energy to come along. One hundred years ago, who could have imagined today's world?

The uncertainties in climate change, however, are not symmetric. The possible (but unknown) downsides are almost unbounded. If we do nothing and get unlucky—temperatures are highly sensitive to carbon dioxide and the harms from temperature increases turn out to be worse than expected—the costs may be terrible. The possible downsides from reducing emissions too much, however, are bounded. We can estimate the cost of replacing our fossil infrastructure with clean energy assuming the worst case, highest possible replacement costs, so we know, within some bounds, what the worst case is for doing too much.[17] That is, if we do too little and we get unlucky (i.e., climate change is far worse than expected) the effects will be very bad. If we do too much and climate change turns out not to be much of a problem, the effects are limited. Moreover, the fossil fuels we will have kept in the ground are still there to be used.

Uncertainty, therefore, should make us more cautious than otherwise. The core uncertainty is that climate change may be far worse than we expect, so we should be willing to spend more to avoid these very bad cases. That is, the conclusions above, which did not consider uncertain but very bad outcomes, were (intentionally) conservative. If anything, we should be doing more than was suggested.

There is also uncertainty about the costs of clean energy technology, as already mentioned. We might invent much less expensive energy sources or, perhaps, ways of capturing atmospheric CO_2 and storing it in a safe place. If we spend resources today to reduce emissions, those expenditures will have been wasted. Perhaps we should not sink costs into reducing emissions when waiting might reduce those costs or reveal more information about the best way to meet our goals.

This uncertainty, however, is counterbalanced by uncertainty about harms. Carbon emissions are effectively permanent, so they are effectively sunk as well. We have the choice of two sunk costs: the costs of clean energy infrastructure or the costs of atmospheric CO_2. Waiting might reveal that either one is cheaper or more expensive than we expect. Whether these offsetting uncertainties make us want to speed up or slow down reductions is hard to say.

The problem of deep uncertainty is challenging and deserves far more attention than I can give it here. I do not, however, think a resolution of the problem would alter the conclusions except possibly to strengthen them, to argue for faster reductions in emissions than I have suggested.

6.6 BARRIERS

The discussion so far has made it seem easy. It is in our collective interest to reduce emissions, drastically, starting now. But it hasn't happened. If it is such a good idea, then we need some explanation for the current state of affairs.

We seem to be hitting ourselves in the head with a hammer. How is that possible?

One explanation is that we are simply being foolish. There is a lot of truth to this. Our foolishness might be explained by disruptions that a serious climate policy could cause. Given that the harms are not immediate, it is always easy to push the pain of emissions reductions back another year. And vested interests can readily take advantage of this. The United States, with its unique political system that creates numerous veto points before action can be taken, is particularly subject to delay and capture by vested interests. Given the vast US emissions, it is difficult for the world to act without the United States.

But even if we overcame US resistance, and the nations of the world were determined to act, the problem would be difficult. It would take another book to explore the problems with reaching an agreement on climate change, but consider some of the problems.

The primary problem is what is known as the "free rider problem." Suppose that the nations of the world are about to agree to reduce emissions, and consider the position of any one nation that is part of the negotiations. If everyone else agrees but that nation pulls out, almost all of the benefits will be reached—all but the one nation will be reducing emissions—but that nation would not have to bear any of the costs. It would free ride on the emissions reductions in the rest of the world. But if each nation has an incentive to pull out of an agreement, it is hard to reach an agreement in the first place. As Benjamin Franklin famously said about the thirteen colonies fighting England, "either we hang together or we hang separately." There is an incentive to let everyone else do the hard work but that might mean disaster.

The incentive to free ride is made even worse because of what is called carbon leakage. A nation that does not join a treaty not only avoids paying for emissions reductions. It can attract high emitting industries relocating from nations that do join the treaty. For example, if the United States were to commit to substantial emissions reductions and China did not, companies might move from the United States to China to avoid the additional costs they would face in the United States. China therefore would obtain two benefits: it would avoid paying for emissions cuts and it would attract industry that might otherwise not locate there. And the same holds for every nation thinking about joining a global climate treaty.

A second, related problem is that nations will bluff to obtain an agreement that is better for them. Reducing emissions will be costly and nations naturally want to bear as little of those costs as possible. Moreover, given that nations are better off free riding than joining a treaty, bluffs are credible. A nation saying "we won't join a treaty unless we get a special deal" might be telling the truth. Bluffs that are not credible are easily ignored but credible bluffs must be taken seriously, making negotiations difficult.

Nations also have very different costs and benefits. Some nations will be hurt badly because of climate change and others not as much. Some nations will bear very high costs of reducing emissions and others not as much. Moreover, nations will value the future differently. A nation that is growing very fast may not want to reduce current growth to obtain future benefits.

Differing costs and benefits normally creates room for bargaining rather than preventing bargaining. If I am going to sell you a widget, you have to value it more than I do for us to agree on a price. But in the context where free riding is a problem, different costs and benefits make free riding easier because it makes it easier to bluff. A nation that just wants to free ride can credibly claim that because of its particular costs and benefits, the treaty does not make it better off.

Finally, all of these incentives play out not just when negotiating a treaty but also when nations have to comply with a treaty. There is an incentive to cheat, free riding on emissions reductions elsewhere without bearing the costs after a treaty is signed—that is, an alternative strategy to refusing to join a treaty is to join and cheat. Monitoring emissions and sanctioning violators is difficult. And nations knowing that enforcement is difficult and that everyone has an incentive to cheat might be even more reluctant to enter into a treaty in the first place.

In sum, just because we know what we need to do does not mean that it will be easy to get nations to agree to do it. Solving the free rider problem in light of the incentives to bluff and to cheat, and given the different costs and benefits each nation faces, will take significant effort, if it is possible at all.

Given these problems, it is especially important, I believe, to pay attention to feasibility constraints. Reaching an agreement that makes each nation better off will be difficult but might be possible. A treaty driven by a theory of ethics that makes major polluters worse off has no chance. They would not be bluffing. They would not even be free riding. They would simply be refusing to take actions that make their citizens worse off.

6.7 CONCLUSIONS

We face tight bounds on our choices regarding climate change. Given the level of past emissions and likely limits on temperature increases, we have used up most of the flexibility that we might have once had. Keeping temperatures below reasonable limits requires transforming our energy system, a process that is likely to be slow because of the sheer size of the system. If we start now and go as quickly as possible, we will still have a hard time keeping temperature increases to a reasonable level. This applies on a global basis. Emissions have to be reduced to zero or near zero in the not too distant future, which means that all countries have to reduce. One of the keys is installing clean energy in the developing world in the first place rather than locking in a fossil fuel infrastructure, which will then either lead to excessive temperature increases or have to be scrapped.

These policy conclusions have important implications for ethical arguments and justice as applied to climate change. They tell us where policies based purely on self-interest—a desire to stop hitting ourselves in the head with a hammer—take us. They take us to aggressive reductions in emissions. The more serious and more immediate the problem of climate change, the more self-interest will lead to aggressive policies. Claims that following our self-interest leads to terrible outcomes or moral corruption do not understand what is in our self-interest.

The outer bounds of reasonable climate policies act also as constraints on ethics. Policies that do not stay within the bounds risk serious harms. In particular, any policy that does not require emissions reductions by all nations in the not-too-distant future will violate basic imperatives to avoid serious harms from climate change. This means that policies like the Berlin Mandate and the Kyoto Protocol should be rejected. If ethical arguments lead us to policies of this sort, they too should be rejected.

Notes

1. David Archer, *The Long Thaw: How Humans Are Changing the Next 100,000 Years of Earth's Climate* (Princeton, NJ: Princeton University Press, 2008).

2. Myles R. Allen, et al., "Warming Caused by Cumulative Carbon Emissions towards the Trillionth Tonne," *Nature* 458 (2009): 1163–1166; Damon Matthews H., Susan Solomon, and Raymond Pierrehumbert, "Cumulative Carbon as a Policy Framework for Achieving Climate Stabilization," *Philosophical Transactions of the Royal Society A: Mathematical, Physical & Engineering Sciences* 370, no. 1974 (2012): 4343–4364; IPCC, "Summary for Policymakers," in *Climate Change 2013: The Physical Science Basis. Contribution of Working Group I to the Fifth Assessment Report of the Intergovernmental Panel on Climate Change*, ed. Thomas Stocker, et al. (Cambridge, UK: Cambridge University Press, 2013), 28.

3. Figure 1 shows the relationship between expected temperature increases and cumulative emissions of carbon, not carbon dioxide. The greenhouse effect comes from the carbon, not the oxygen in CO_2. To translate carbon to CO_2, multiply by 3.67, which is how much more CO_2 weighs than carbon per unit of carbon.

4. We might find a way to capture some emissions and store them safely, which would mean we could continue to emit to this extent. Net emissions would have to be zero.

5. James Zachos, et al., "Trends, Rhythms, and Aberrations in Global Climate 65 Ma to Present," *Science* 292, no. 5517 (2001): 686–693.

6. I ignore the possibility of attempting to manipulate the atmosphere to offset the increased warming, for example, by putting reflective aerosols in the atmosphere to reflect

sunlight away from the earth, thereby lowering temperatures. This process, sometimes called geoengineering,
is extremely risky and would not undo all of the effects
of climate change. Naomi E. Vaughan and Timothy M.
Lenton, "A Review of Climate Geoengineering Proposals,"
Climatic Change 109, no. 3–4 (2011): 745–790. We may
also never be able to get to zero emissions because some
emissions, such as nitrous oxide and methane emissions, may be unavoidable, in which case we need to
integrate emissions floors into our planning. For a discussion, see, Niel H. A. Bowerman, et al., "Cumulative
Carbon Emissions, Emissions Floors and Short-Term
Rates of Warming: Implications for Policy," *Philosophical
Transactions of the Royal Society A: Mathematical, Physical
and Engineering Sciences* 369, no. 1934 (January 13,
2011): 45–66.

7. Trillionthtonne.org, Oxford e-Research Centre, Department
of Physics, University of Oxford, http://trillionthtonne.org/

8. Mark New, et al., "Four Degrees and beyond: The Potential
for a Global Temperature Increase of Four Degrees and Its
Implications," *Philosophical Transactions of the Royal Society
A: Mathematical, Physical and Engineering Sciences* 369, no.
1934 (2011): 6–19.

9. There is a rich literature on whether the link between
energy and income has been decoupled because of the transition of modern economies to services. See, for example,
K. Bithas and P. Kalimeris, "Re-Estimating the Decoupling
Effect: Is There an Actual Transition toward a Less Energy-
Intensive Economy?," *Energy* 51 (2013): 78–84. Even in
the most optimist version of the decoupling hypothesis,
energy use still grows with income, only at a slower rate
than previously.

10. International Energy Agency, "World Energy Outlook"
(Paris: International Energy Agency, 2013). https://www.iea.
org/publications/freepublications/publication/WEO2011_
WEB.pdf.

11. Vaclav Smil, *Energy Myths and Realities: Bringing Science to the
Energy Policy Debate* (Lanham, MD: Rowman and Littlefield,

2010; Vaclav Smil, *Energy Transitions* (Santa Barbara, CA: Praeger, 2010).

12. International Energy Agency, "World Energy Outlook" (Paris: International Energy Agency, 2011), 231–232.

13. Intergovernmental Panel on Climate Change, *Climate Change 2007—Mitigation of Climate Change, Contribution of Working Group III to the Fourth Assessment Report of the IPCC*, eds. Ben Metz, et al. (Cambridge, UK: Cambridge University Press, 2007), 265. Reports that reserves are 12,800 $GtCO_2$ which translates to about 3.5 trillion tons of carbon.

14. The following paragraphs crib liberally from Dieter Helm, *The Carbon Crunch: How We're Getting Climate Change Wrong— and How to Fix It* (Yale University Press, 2013), 40–48.

15. Helm, *Carbon Crunch*, 42.

16. Yang Ailun and Yiyun Cui, "Global Coal Risk Assessment, Data Analysis and Market Research," Working Paper, WRI Working Paper Series (Washington DC: World Resources Institute, 2012). http://pdf.wri.org/global_coal_risk_assessment.pdf.

17. There are a large number and variety of engineering studies that make these sorts of estimates. See for example, Mark Z. Jacobson and Mark A. Delucchi, "Providing All Global Energy with Wind, Water, and Solar Power, Part I: Technologies, Energy Resources, Quantities and Areas of Infrastructure, and Materials," *Energy Policy* 39, no. 3 (2011): 1154–1169; Mark A. Delucchi and Mark Z. Jacobson, "Providing All Global Energy with Wind, Water, and Solar Power, Part II: Reliability, System and Transmission Costs, and Policies," *Energy Policy* 39, no. 3 (2011): 1170–1190.

7

The Role of Claims
of Justice in Climate
Change Policy

PHILOSOPHERS HAVE APPLIED theories of justice to determine what should be done about climate change. The three core types of theories they have applied to climate change are theories of distributive justice, theories of corrective justice, and theories based on equality.[1] I will examine each of these types of theories in this chapter and show that, as they have been applied to climate change, they all suffer from two faults: they suffer from what I call climate change blinders and they fail basic tests of feasibility.[2]

There are many versions, complexities, and details of each of these classes of theories. At the risk of doing great injustice to theories of justice, I will focus on core elements of each of these three classes of theories, and particularly on those elements that have been used to make arguments about climate change.

Without discussing each and every version of every possible theory that might apply, I cannot claim that theories of justice have nothing to say about climate change policy. There could always be some theory or an alternative version of a theory that does have something to say. I can only examine what has been argued so far, and in a short monograph, can only cover the central arguments, not each nuance. The best I can say is that this invites a response explaining why a particular theory or element of a theory plausibly changes my conclusions.

7.1 DISTRIBUTIVE JUSTICE

Theories of distributive justice are concerned with the distribution of benefits and burdens in a society. There are many theories of distributive justice including utilitarianism, Rawls's principles of justice, theories of equality of opportunity, and the capabilities theory.[3] They vary in terms of which members of the population they focus on; their measure of well-being; how they balance different elements of well-being such as income, health, and dignity; and the extent to which they consider factors such as luck and opportunity in addition to, or as separate from, well-being.

While acknowledging important differences, for our purposes, we can crudely lump theories of distributive justice together as theories focused on the distribution of well-being. These theories argue that actions or policies that help the badly off because they are badly off, however defined, are required as a matter of justice.

Distributive justice arguments have been applied in the climate change context to conclude that wealthy nations should bear most, or all, of the costs of reducing emissions. Asking poor nations to reduce or not increase emissions condemns them to staying poor and ignores the vastly unequal distribution of resources. How can we ask

300 million rural poor in India to continue to live without electric power when Americans are choosing to live in massive air-conditioned homes, driving SUVs through the suburban sprawl to play golf on grass in the dessert or to swim in heated pools in Chicago? Any reasonably fair approach to climate change would ask that the wealthy sacrifice first and sacrifice more.

The argument was put forth forcefully by Henry Shue (who uses methane from ruminants as his example, but the point is completely general):

> The central point about equity is that it is not equitable to ask some people to surrender necessities so that other people can retain luxuries. It would be unfair to the point of being outrageous to ask that some (poor) people spend more on better feed for their ruminants in order to reduce methane emissions so that other (affluent) people do not have to pay more for steak from less crowded feedlots in order to reduce their methane and nitrous oxide emissions, even if less crowded feedlots for fattening luxury beef for the affluent would cost considerably more than a better quality of feed for maintaining the subsistence herds of the poor.[4]

Paul Harris puts the point as follows:

> One thing that seems unassailable from the perspective of world ethics and global justice is that greenhouse gas emissions required for subsistence take priority over other kinds of emissions, and ought not be subject to any kind of limit. . . . This means that the 'luxury' emissions of rich people, regardless of where they live, ought to be a focus of greenhouse gas cuts. . . . Arguing the point is as good as saying that some Rwandans should die so that some Virgin Islanders can recharge their mobile phones.[5]

Stephen Gardiner argues that "there seems to be a broad ethical consensus that developed countries should shoulder most of the burden of action (at least initially), and so should have fewer emissions."[6] He argues that this conclusion is supported by many theories of justice including distributive justice. He notes that "theories that prioritize the interests of the least well off endorse the consensus [i.e., the conclusion quoted above] because the developing countries are much poorer than the developed countries." Numerous other authors adopt similar approaches.[7]

Shue's statement highlights a key feature of how theories of distributive justice apply in the climate change context. He would increase the cost of emissions reductions in the name of justice. The feedlot changes for luxury beef cost considerably more than the feed changes for the subsistence ruminants. Nevertheless, we should pursue the feedlot changes for luxury beef to promote justice.

This is a general feature of policies based on distributive justice. They allocate obligations to reduce emissions based on distributive concerns, which means that they do not seek to find the lowest cost reductions. Unlike Shue, most authors do not highlight this explicitly, but I take it to be implicit in their arguments.

Some commentators miss this key feature of distributive justice. For example, James Garvey uses a claim that rich nations have greater capacity to reduce emissions to support the argument that distributive justice demands that the rich do more.[8] If the basis of the obligations to reduce emissions is cost effectiveness and capacity, however, we can put the philosophy texts away and ask the economists and engineers to find the right solution. Distributive justice is doing no work. It is nice if there is less inequality when we pursue the efficient solution because it means that the baseline distributive problems are lower, but as long as there is inequality, distributive justice demands that we deviate from the cost-effective solution to help the poor. That is, suppose that pure efficiency concerns imply that developed countries go first and do more than developing countries. When we add distributive concerns, developed countries would have to do even more still because the distributive gains would be worth the efficiency losses. I will follow Shue and take distributive justice to imply that we spend more to stop climate change than otherwise by allocating emissions reductions based on distributive concerns, not just efficiency.[9]

7.1.1 Climate Change Blinders

Suppose we agree, based on a theory of distributive justice, that people living in wealthy countries have an obligation to transfer another $100 to (poor) people living poor countries. (The obligation could be any amount including very large sums; $100 is a placeholder for an arbitrary number.) To fulfill this obligation, the people living in wealthy countries might send cash to the governments of poor countries, but this may not be the best course of action. There may be little reason to believe that those governments will spend the cash in a way that will help their citizens rather than themselves. People in wealthy countries may instead provide cash conditional on certain behaviors or only send cash to places where it is likely to be well spent.[10] They may alternatively send specific goods or services. For example, they may provide free insecticide-treated mosquito nets, vaccines, medicines, schools and education supplies, technology, micro-credit loans, fertilizer, or any number of other things. Wealthy countries might also change their trade, patent, farming, immigration, or other policies in ways that hurt themselves but help poor people living in poor countries by $100. Or wealthy countries might agree as part of a climate treaty to bear $100 more of the burden of emissions reductions than they would otherwise agree to.

The question is whether distributive justice demands that this last action be taken, that a climate treaty be the mechanism for transferring the $100 to the poor instead of any of the other actions. One could of course argue that all of these actions should be taken cumulatively so that even more is done to help the poor, but remember, we've already used our theory of distributive justice to conclude that the obligation wealthy nations or people owe to the poor is $100. You can change the $100 number to be whatever you want. The argument does not depend on any particular view on the size of the obligations of the rich to the poor. It could be $100 billion in total, $100 trillion, or more. (To get a sense of scale, the United States currently gives about $20 billion per year in foreign aid, although much of that is not to help the poor.) The only assumption is that there is some obligation stemming from justice and transferring more than that is not required.

Does distributive justice demand that the transfer of $100 to the poor be done via climate change policy? The answer is straightforward: distributive justice allows, and arguably even demands, that wealthy countries choose the option, or set of options, that transfers the resources most effectively. If the amount poor people are to receive is $100, we should want wealthy countries to choose the options for transferring the $100 that are most effective.

The reason is that theories of distributive justice care about well-being. Given a choice between, say, changing trade policies, cash transfers, mosquito nets, fertilizer, and designing a climate change treaty, distributive justice demands a choice based on which combination best promotes well-being. Choosing a less effective policy reduces overall well-being by reducing available resources. If there are two policies, one which costs $130 and one which costs $120, and both transfer $100 to the poor, distributive justice demands that we choose the policy that costs $120 as this best promotes overall well-being. By choosing the policy that costs only $120, there is an additional $10 of resources that can be shared or perhaps allocated to the poor.

At this point, the work of philosophers and of theories of distributive justice is done. The choice of which policies or combination of policies is most effective is a matter of economics, political science, international relations, and related fields that try to measure and describe the effectiveness of policies. Theories of justice are not going to tell us whether reducing subsidies for farmers in wealthy countries is a better way of helping the poor than setting emissions reduction goals with distributive effects in mind, or whether debt forgiveness, patent policy, technology transfers, vaccines, micro-credit, trade policy, or some combination is better still.

Theories that argue that we must design a climate policy based on distributive justice give a different and wrong answer. They require that wealthy nations make transfers to poor nations within the context of climate policy rather than considering how best to make transfers within the entire set of policies. They see a problem in front of us to be solved—climate change—and assume that the solution must take distributive considerations into account. They operate with climate change blinders. When we take off the blinders, we see that we have two serious problems: (i) climate change and (ii) a large number of people living in poverty at a time when others are enormously wealthy. We need to solve both. But we do not necessarily need to solve them both with the same policy. Philosophical theories that demand we use the same tool ignore the wider set of tools that are available and that might solve these critical problems better.

Some theories of distributive justice, however, demand a just distribution of particular goods. We might not be able to substitute for a redistributive climate treaty the combination of a nonredistributive climate treaty and a policy that transfers a different good to the less well off. John Rawls, for example, focuses on a particular set of goods he called primary goods. Amartya Sen and Martha Nussbaum focus on a list of capabilities. Others focus on the allocation of opportunity. Giving someone money, micro-credit, or a mosquito net may not substitute for the unequal distribution of a good that is central to justice or to human dignity. We cannot take away someone's dignity or self-respect and then simply compensate them with cash. Certain goods, many theorists believe, are incommensurable.

Perhaps climate change is like this. Perhaps we cannot simply enter into a climate treaty based on cost-effectiveness considerations or raw self-interest and then make it up elsewhere, with, say, a change in farming or patent policies.

Henry Shue's argument about the obligation to contribute to the creation of a public good hints at this.[11] He proposes that "[a]mong a number of parties, all of whom are bound to contribute to some common endeavor, the parties who have the most resources normally should contribute the most to the endeavor." Some have read this statement to imply that the funding for each particular good must take distributive concerns into account, with no possibility of trading off across policies.[12]

The strongest version of the incommensurability argument is that some level of emissions are needed for subsistence. Denying someone the minimum emissions needed for subsistence would cause him great harm. This minimum cannot be traded off for other goods. A related claim is that people and nations have a right to develop. Energy is needed to develop. A climate policy that limits emissions in poor countries might prevent development. It would force poor people to stay poor, and, therefore, does not meet the demands of distributive justice.

While climate change policy is about energy, it concerns the type of fuel we use for energy, not whether we use energy. Clean energy is more expensive than fossil fuel energy, so the problem at its core is about money, not about incommensurable goods, basic needs, dignity, primary goods, capabilities, or some other requirement of justice.

Once we are transferring money to the poor, we need to think about the best way to do that. It might be that allowing the poor to use fossil fuels and requiring the wealthy to use renewable energy is the best way, but this is not a matter philosophy or philosophers are equipped to answer. The problem is complex. Renewables are not as reliable as fossil fuels and because they tend to be intermittent, do not work well for what is called base load power. There are, right now, no good renewable energy sources for transportation.

Nuclear power comes with its own trade-offs, particularly in unstable countries. But building new, long-lived fossil fuel infrastructure where it does not now exist effectively locks in emissions for the long-term. The design of energy supply systems is complex.

We also need to compare the trade-offs in the use of renewables with other uses of resources, such as spending on education, the legal system, and so forth. These are hard policy problems and it is important to try to get them right. It might be the case that adding new fossil fuel infrastructure in developing nations is the best approach, all things considered. There is nothing, however, in the requirement that we pursue distributive justice that informs these choices.

The better version of the incommensurability argument is that if climate policies are insufficiently aggressive at reducing emissions, the harms cannot be compensated. People's ways of life might be altered due to the resulting climate change. Coastal areas might be flooded, forcing migration. Ocean acidification and increases in water temperatures may force people reliant on the oceans to find other ways of living. Weather patterns may change so that agricultural productivity may decline in many areas, again forcing people to change how or where they live. These changes may be incommensurable with offsetting transfers. Giving tens of millions of Bangladeshis money to relocate from their flooded homes may not meet the demands of justice. Their homes have been destroyed.

If one believes this incommensurability argument, that the ways of living that will be lost because of climate change are inviolate, then the most important thing to do is to stop climate change. Cost-effectiveness becomes even more important rather than less. We even less want to alter climate policy to allow for distributive considerations because the most important way to achieve distributive justice is to stop climate change.

A related argument is that climate change will make the poor worse off. A treaty to address climate change cannot be treated as separate from distributive concerns because one of the very purposes of the treaty is to prevent the poor from suffering. As Darrell Moellendorf states, "When climate change is likely to throw people into desperate poverty and set back human development in some states that are making progress, a treaty that seeks to prevent these evils is not addressing matters of distributive justice that are external to concerns about climate change."[13] Moellendorf concludes that distributive concerns are central to climate change policy.

This is correct but strengthens rather than weakens the claims made here. Because climate change may make inequality and poverty worse, it is especially important to prevent it. An effective climate treaty is especially important. But this means that choosing more expensive climate policies such as the feedlot changes Shue recommends is an even worse idea.

7.1.2 Feasibility

Not only do applications of theories of distributive justice suffer from basic logical flaws. They violate feasibility constraints. They do so in two ways. The first is that they do not offer a path to stabilizing emissions. As chapter 6 discussed, we cannot allow substantial increases in the fossil fuel capacity anywhere and still meet reasonable climate goals. Approaches like the Kyoto Protocol that exempt large portions of the world from obligations to reduce emissions are not a serious option regardless of their justice-related merits. If this is what distributive justice demands, we should reject distributive justice.

More modest versions of the argument would allow developing countries to delay their emissions reductions while demanding that developed countries begin immediately. We may still be able to stabilize the climate with this approach, but at either higher levels of CO_2 concentrations and temperatures or at higher costs. Achieving distributive goals almost always involves costs, so perhaps these costs are worth absorbing.

The room for this sort of flexibility, however, is limited. If developing nations install substantial new fossil fuel infrastructure, they will have locked in emissions for the life of that infrastructure unless they are willing to scrap it, an unlikely possibility. Just installing the fossil fuel power plants already in the planning stage will mean that we cannot meet the 2°C goal. Adding additional fossil fuel infrastructure all but guarantees bad outcomes. The needed limits on greenhouse gases tightly constrain the room for poor nations to increase their emissions.

The world has recognized this feasibility problem. The dichotomous distinction approach taken in Kyoto has now been replaced with the Durban approach, which requires all nations to reduce emissions. The Durban approach has not yet, as I write, been translated into actual targets, but if we are to achieve reasonable emissions reductions goals, it is the only feasible approach. Suggesting a return to the Kyoto approach is simply a suggestion for failing to meet basic climate change goals.

Philosophers such as Gardiner now recognize the feasibility problem. As recently as 2004, Gardiner endorsed the approach taken in the Kyoto Protocol, arguing that "there is a surprising convergence of philosophical writers on the subject: they are virtually unanimous in their conclusion that the developed countries should take the lead in bearing the costs of climate change, while the less developed countries should be allowed to increase emissions for the foreseeable future."[14] He now agrees that this approach is not a feasible path to stabilization and that developing nations also must reduce emissions: "[i]t is now clear that the developing nations will have to constrain their own emission quickly and significantly . . . From the point of view of justice, the consensus appears to be that this only increases the burden on developing [*sic*] countries to assist in other ways."[15] If these other ways include the full range of policies that might be used to help the poor, then Gardiner and I agree on distributive justice. Feasibility binds—that is why all nations have to constrain emissions quickly—and we have to consider other ways of helping the poor, with no climate change blinders.

The second feasibility problem, which is to great extent a reflection of the first, is that many wealthy nations will not agree to a treaty that requires them to reduce emissions if fast-growing developing nations do not also agree to reduce, or at least not increase, emissions. Why reduce emissions if there will be no long-term climate benefit because developing nations are allowed to increase emissions? As the unanimous Byrd-Hagel resolution demonstrates, the United States, and likely many other developed countries, would view such a treaty as far outside the bounds of their self-interest. The costs would be greater than the climate benefits.

Perhaps, one might argue, the United States and other developed nations should enter into such a treaty anyway, because it is required by justice. But asking the United States and other developed nations to voluntarily enter into a treaty that makes them worse off is not a feasible approach to solving the problem of climate change.

Overall, the arguments that a climate treaty should be based on distributive justice fail. They are based on a vision that wears climate change blinders, one that fails to consider the myriad ways that obligations stemming from distributive justice can be met. They often produce climate treaties that by their design fail to achieve the necessary reductions. Thankfully, in the most recent negotiations in Durban, the world has begun to abandon this approach in favor of alternatives that have a better chance of reducing emissions at a reasonable cost.

7.2 CORRECTIVE JUSTICE

Arguments based on theories of corrective justice view emissions of CO_2 as a wrongful act and demand that emitters, whether nations or individuals, make compensatory payments to those who are harmed by these emissions. Nations or people that have historically been high emitters—largely, but by no means exclusively, developed nations and their residents—would owe compensation to nations or people who have historically been low emitters. These claims are rooted in theories of responsibility for actions, theories that go back to Aristotle. The Pottery Barn motto—you broke it, you own it—captures the intuition.

Peter Singer encapsulated the intuitions succinctly:

> To put it in terms a child could understand, as far as the atmosphere is concerned, the developed countries broke it. If we believe that people should contribute to fixing something in proportion to their responsibility for breaking it, then the developed nations owe it to the rest of the world to fix the problem with the atmosphere.[16]

Many other commentators endorse this approach.[17] Henry Shue treats it as his first principle of equity:

> When a party has in the past taken an unfair advantage of others by imposing costs upon them without their consent, those who have been unilaterally put at a disadvantage are entitled to demand that in the future the offending party shoulder burdens that are unequal at least to the extent of the unfair advantage previously taken, in order to restore equality.[18]

After reviewing the arguments concerning responsibility, Stephen Gardiner concurs:

> The arguments in favor of ignoring past emission are, then, unconvincing. Hence, contrary to many writers on this subject, I conclude that we should not ignore the presumption that past emissions pose an issue of justice that is both practically and theoretically important.[19]

There is also a legal basis for this approach. The Framework Convention in its introduction notes:

> That the largest share of historical and current global emissions of greenhouse gases has originated in developed countries, that per capita emissions in developing countries are

still relatively low, and that the share of global emissions originating in developing countries will grow to meet their social and development needs.

More generally, some countries use what is known as the "polluters pay" principle, at least in some circumstances. It means what it says: polluters must pay for any harms that they cause. This is by no means universal and in its most general form is largely unknown because most countries make polluters pay only if they are negligent, except in very narrow circumstances. Even if honored in the breach, the polluters pay principle is often seen as a widely accepted principle. It is thought to be based in corrective justice and also to promote efficiency because it forces actors to take the full costs of their actions into account.

In prior work, Eric Posner and I discussed a number of problems with applying the corrective justice claim on its own terms.[20] One is that corrective justice normally applies only to wrongful acts, meaning that the person who did the act was negligent or otherwise acted with fault. People who emitted CO_2 before the problem of climate change became well known were probably not negligent. They could not have known that heating their homes or driving to work hurt others.

Once the problem of climate change became known, perhaps people engaging in these acts might have been acting negligently. But people find themselves in widely differing circumstances. Some people live in cold climates with few renewables and may have few choices other than to use fossil fuel energy to supply heat. Others live in places with abundant renewable energy or in temperate climates and may not need fossil fuels. Others live in large nations and need fuel for transportation. Attributing fault requires complex judgments about what behavior is acceptable. Is it negligent to live somewhere cold?

A second problem is that theories of corrective justice were developed to apply to individuals, not groups. When we try to apply it to groups, there is a concern that the wrong people will be made to pay and the wrong people will be compensated. Collective responsibility is often viewed as immoral. There is, however, no way to identify wrongful emitters and their victims on an individual basis. Instead, the actors in a climate treaty are nations.

Finally, many wrongful emissions were by people who are no longer alive or who are old enough that they will not bear the costs of emissions reductions. There is no way to make these individuals pay. Making current people pay imposes obligations based on corrective justice on people who did not act wrongfully.

The standard response to the latter problem is to apply corrective justice at the national level and to let the internal allocation of responsibility be determined by each nation.[21] The United States, for example, would owe some amount based on the past emissions of people who live there, and it could determine how to allocate the costs to its residents. The United States has persisted over time and might be viewed as the responsible agent. Even though current citizens could not have wrongfully emitted CO_2 before they were alive, they most likely received some of the benefits of those emissions.

The benefit theory is not the same as the polluter's pay or wrongdoer theory. It is a theory of ill-gotten gains rather than past wrongs. The amount owed would be different, relating to the gains rather than the wrongs. It is a much more tenuous theory and has been criticized by a number of authors.[22] Perhaps limiting responsibility to emissions after some date, such as 1990, solves both the intergenerational problem and the problem with determining fault, but there are arguments that it does not.

These sorts of arguments have been discussed in the literature, and a full exploration of applying corrective justice to climate change would require grappling with them. Even if we could work out these details, however, there are larger problems. The corrective justice argument suffers from the problems of climate change blinders and feasibility.

7.2.1 Climate Change Blinders

To understand the blinders problem, we first need to clarify the logic of the corrective justice claim. The claim is not about the harms from the climate change that have happened so far. In the last century, temperatures have increased a little less than 1°C. While there likely have been harms from this increase, they have been mild and hard to distinguish from harms due to natural variations in the weather. If the demand were for compensation for these harms, the amounts would not be large and would not significantly affect the allocation of obligations under a climate change agreement.

The real claim lies elsewhere. It is that some nations or individuals have used more than their share of the limited ability of the atmosphere to absorb CO_2. As noted in chapter 6, the atmosphere has a fixed capacity to absorb CO_2. A large portion of it has been used by one set of nations or individuals. This means that other nations or individuals cannot use that capacity. The "you broke it" claim is that you used up a limited resource, not that the climate in 2015 is already broken in the sense of causing terrible harms. For example, if the capacity of the atmosphere to absorb carbon is one trillion tons before the harm becomes too large, we have used up more than half of that already. The corrective justice claim is that this use of the limited capacity of the atmosphere creates an obligation to compensate because it violates rights or is otherwise unjust.

Once we see that this is the structure of the corrective justice argument, the blinders argument is clear. The corrective justice argument applies to any limited resource anywhere in the globe that is not shared equally. Anytime one nation or set of individuals uses more than their share of a limited resource, others are harmed because they cannot use that resource, and, therefore, those who exploited the resource would owe compensation to others.

Few if any resources are shared equally. Nations and individuals have differing amounts of land, good agricultural soils, and fresh water. Some nations have vast oil, gas, or coal reserves. Others have forests, rare minerals, or diamonds and gold. Some are warm, others brutally cold. Some have few resources. Others are have many. Any unequal sharing of scarce resources would create an obligation under corrective justice using precisely the same logic as for the atmosphere. One can take the very same sentences and substitute in any of these the resources as long as the resource is reasonably fixed in supply and important to people. Peter Singer, for example, states:

> If we begin by asking, "Why should anyone have a greater claim to part of the global atmospheric sink than any other?" then the first and simplest response is "No reason at all." . . . This kind of equality seems self-evidently fair . . .[23]

Now substitute for "global atmospheric sink" minerals, oil, forests, fresh water, fertile soil, or any other limited resource. Perhaps for some there might be some reason for unequal sharing of some resources, but for most the argument is identical. There seems to be no reason at all why some humans should have greater rights to valuable resources merely by virtue of their birth.

The problem goes to the core of the corrective justice argument. If you believe the corrective justice argument, you either have to be able to distinguish these other cases or be willing to accept the conclusion of the argument and suggest that all resources be shared equally. Anytime anyone uses more than their equal share of a limited resource, they owe compensation under the corrective justice argument.

Most authors make the corrective justice claim without recognizing its scope. They have no general theory of resource allocation. They operate with climate change blinders. They are thinking only about climate change and not the general logic of their arguments.

If we remove the blinders and then try to find a way to apply the corrective justice argument only to climate change, we see that it is difficult or impossible. Consider some possibilities.

One possibility is to invoke a principle of territory or physical possession. If you are on top of the resource— the forest is on your sovereign land, the oil is under your ground and so forth—it is not a violation of corrective justice to exploit more than a pro rata portion of it. A related possibility would be to invoke international norms or international law, which say that a nation owns resources within its territory. The atmosphere is not solely within any nation's territory, so perhaps, like the sea, it is owned by all. This, however, assumes its conclusion. We would need a reason, based on a theory of justice, why territorial possession is fine while possession through use (as the high emitting countries have done to the atmosphere) is not.

A territorial theory could not be based on the claim that existing territorial boundaries were set on a just basis because many were not. Many territories were acquired by force and many simply by use. There are few places on Earth where territories were based on a claim of justice.

Another alternative might be a principle of legacy. Maybe if you have possessed something long enough, theories of corrective justice will not question the validity of owning it. Current residents justly possess all of the Americas even though their ancestors unjustly took it from prior possessors by force. The ownership is just because the injustice is old. Corrective justice would then apply differently to the atmosphere because rich nations took an unfair portion of it only recently. This argument, however, seems untenable. It is hard to see why old wrongs are better than new wrongs. Moreover, it runs directly into the problem that people living today are not the same as the people who emitted in the past. If we are going to use a collective theory of responsibility to claim that current people are responsible for emissions of their ancestors, the same collective theory would apply to other resources.

Peter Singer is one of the few philosophers who recognizes that there is a climate change blinders problem with the corrective justice argument. He devotes a number of pages to distinguishing the atmosphere from property rights in other resources.[24] He relies on the argument for private property put forth by John Locke. Locke argued that private property is just only if, when one claims it, there is enough (or possibly more) left for others (say because you use the property to produce goods which improve overall welfare). Singer says that this is not true of the atmosphere: if one group of people uses it as a sink to dispose of their waste, others cannot. It is a limited resource.

I've always found Locke's argument mysterious, at least as applied to any good that is scarce (that is, almost everything). Although Singer is right that the atmosphere is a limited resource, the argument does not distinguish the atmosphere from any other limited resource, including the classic case of private property, land. There are limited amounts of good land, and ownership by a select few may not leave enough for others. Many people are forced to cultivate land that is dry or infertile. And it is definitely not true of minerals and of numerous other valuable resources. For example, some countries have an adequate supply of fresh water but many do not. The territorial claims to fresh water by the lucky few mean that the rest have too little.

Some advocates of the corrective justice claim try to dismiss the problem as theoretical nitpicking. For example, Stephen Gardiner argued, we do not need a "complete background understanding of international justice" to (my words, not his) realize that the rich countries' seizure of the atmosphere is unjust.[25] We can fix the problem with unfair use of the atmosphere without fixing the unequal distribution of other resources or trying to correct every other type of injustice. Let us just deal with the problem at hand rather than waiting for a complete theory of everything.

This response strikes me as completely inadequate. If your theory of justice has parts which you like (compensation for use of the atmosphere) and parts you don't like (compensation for unequal use of all other limited resources), you can't just take the parts you like. Not without some theory or explanation. And especially if the parts you don't like are larger, by several orders of magnitude, than the parts you do like. (Compensation for all other unequally shared resources probably swamps compensation for past unequal use of the atmosphere.) It is not as if the problem is one of some minutiae with the theory not yet being fully worked out, a few i's not yet dotted and t's crossed. The problem is that the advocates of the theory are unwilling to accept the most important implications of their theory and cannot, or are unwilling to, explain why except to wave their hands. What is philosophy for if not to try to think through problems carefully?

In fact without distinguishing other unequally shared resources, we would not even know if payments for use of the atmosphere on net go in the right direction. With dozens or hundreds of cross-cutting claims for unequal use of resources, a nation or individual who is actually a net wrongdoer might receive payments if we base them only on CO_2 emissions. Nations that should be net recipients might have to instead make payments.

The most plausible argument for dismissing claims regarding unequal use of resources other than the atmosphere is that equal sharing of all resources is hopelessly infeasible. We are not going to give back the United States to Native Americans. South Africa is going to keep its diamonds, Russia its gas, and Saudi Arabia its oil. Canada's weather is unfairly cold and Egypt and Libya don't enough rain. Iceland is blessed with geothermal energy; Japan has no local energy sources. Suggesting that we fix these inequalities sounds almost wacky because it is so infeasible. But we can negotiate a climate treaty that provides compensation.

Peter Singer ultimately resorts to an argument of this sort, that we should let bygones be bygones. He concludes:

> The putatively historical grounds for justifying private property put forward by its most philosophically significant defenders—writing at a time when capitalism was only beginning its rise to dominance over the word's economy—cannot apply to the current use of the atmosphere ... [The arguments for private property] imply that this appropriation of a resource once common to all humankind is not justifiable.[26]

In other words, he cannot distinguish other types of resources but that is no excuse for not allocating the atmosphere fairly. The argument is instead one of feasibility. Let us turn then to feasibility.

7.2.2 Feasibility

Feasibility arguments are a slim reed on which to base corrective justice claims. Even a cursory examination of the feasibility of compensation in the climate context shows that it is as unrealistic as asking for compensation for the unequal allocation of oil, diamonds, rare minerals, forests and most other resources.

A back of the envelope calculation shows the problem. We want to estimate the payments that would be due under a corrective justice theory. Suppose that each nation has an obligation to make a payment or has the right to receive a payment based on any difference for any year between its per capita emissions and the global average per capita emissions during that year. The payment would be the excess use multiplied by a price per ton of CO_2, and the receipt would be the opposite. For example, if the global average of emissions in a year is one ton per person, and a nation emitted 1.5 tons per person, the nation would owe compensation based on the excess half ton per person, multiplied by the number of people living there and the price per ton. If a nation had emissions lower than the global average it would similarly have a right to receive compensation. We would do this for each nation and each year to arrive at a total amount.

If we perform this calculation going back to 1850 and if the price of CO_2 is \$35/ton (which is the price currently used in the United States when making climate change policy), the United States would owe just under \$12.4 trillion (that's \$12,400,000,000,000).[27] The United States currently gives about \$20 billion in foreign aid, so it would have to agree to pay more than six hundred years of foreign aid as part of the proposed climate treaty.

The numbers are similarly daunting for other nations. The EU would owe \$4.2 trillion. India would have a right to receive \$10.8 trillion and China would have a right to receive \$9.9 trillion. Brazil would owe \$1.6 trillion because of its massive deforestation. Russia would owe \$3.4 trillion. Transfers of this size are not likely now, or ever. Feasibility cannot distinguish the climate from many other unevenly shared resources.

The response might be that poor nations will not agree to a climate treaty unless it is fair, so in fact feasibility concerns cut the other way: only a treaty that addresses the prior unjust use of the atmosphere is feasible.[28] Crudely, India could say that it will not agree to a climate treaty unless it receives the payment of the \$10.8 trillion that is owed.

This possibility is borne out in what are known as ultimatum games. In these games, one person is given an amount of money, say $10. He can keep as much as he likes and give the rest to a second person. The second person has the choice of accepting the money or refusing it. If the second person refuses it, nobody gets anything. From a purely rational standpoint, the second person is better off even if he only gets one penny and the first person keeps $9.99. He ought to accept any amount other than zero because he will get something rather than nothing. What actually happens is that the second person most often rejects the offer unless the split is reasonably fair. Poor nations may similarly reject a climate treaty that is otherwise in their interest if they do not perceive it to be fair.

If this is the case, then we are in a very bad spot because payments of the sort demanded by corrective justice are not going to be forthcoming. Rich nations will not agree to payments of this magnitude, particularly if the reason is that poor nations are demanding the payment to agree to a climate treaty that is otherwise in their own interest. The resulting stand-off would mean that emissions go uncontrolled, hurting everyone. Poor nations would suffer from climate change and will not get the payments they hope for.

I am hopeful that this will not come to pass. Nations are made better off by reducing emissions. Poor nations, those most likely to receive corrective justice payments, may in particular stand to benefit from global emissions reductions. If these nations make demands for additional payments that prevent the world from reaching a climate treaty, they hurt themselves more than they hurt others.

To recap, the internal logic of the corrective justice claim only holds together on feasibility grounds. This is the only way to distinguish the atmosphere from other resources, as commentators such as Peter Singer recognize. If a treaty based on corrective justice is not feasible, the theory fails. But such a treaty is not remotely feasible.

7.3 EQUAL PER CAPITA EMISSIONS

The final claim about justice that I will examine is the claim stemming from equality. The claim is that justice requires that each individual receive an equal share in the remaining capacity of the atmosphere to absorb emissions. For example, if the safe cumulative emissions limit is 1 trillion tons of carbon and we have emitted around 600 billion tons, there are 400 billion tons of remaining capacity. There are 7.125 billion people alive today. Each person would be allowed their share, which is 600 billion tons divided by 7.125 billion people or just 60 tons of carbon per person.

Equal per capita emissions rights are often envisioned as embedded in a cap-and-trade regime. Each person would get the right to emit about sixty tons. Instead of using that right, they could sell it and allow the buyer to emit some, or all, of the 60 tons. Industries in countries with large fossil fuel infrastructures, like the United States and Europe, could buy emission rights from those without larger infrastructures, ensuring that emissions would occur where needed based on cost considerations while individuals would share equally in the atmosphere.

In addition, because it might be infeasible to give emissions rights to all seven billion individuals directly, equal per capita rights regimes often contemplate giving the rights to countries on the basis of population. The governments would then determine how to use the emissions rights.

The appeal of an equal per capita rights system is based on a view that the atmosphere is a common resource. It belongs to everyone. When we divide it up, therefore, everyone should get an equal share. As Paul Baer noted, "[t]he central argument for equal per capita rights is that the atmosphere is a global commons, whose use and preservation are essential to human well-being."[29] Baer has also stated, "Ethically, disparate claims to common resources are difficult to justify."[30] Because of its simplicity and intuitive appeal, equal per capita rights has been described as "the most politically prominent contender for any specific global formula for long-term allocations with increasing numbers of adherents in both developed and developing countries."[31]

The intuition is similar to the arguments for corrective justice. Equal per capita rights claims and corrective justice claims are both based on a view that the just allocation of the atmosphere is pro rata. The corrective justice claim is that people owe compensation for more than pro rata use in the past. The equal shares claim is that people should be given a pro rata share of what remains for future use.

The equal shares argument is stronger than the corrective justice argument because it does not have to deal with many of the problems that plague the corrective justice: deciding who acted wrongly in the recent and distant past, tracing those wrong acts to people alive today who can be held responsible, applying responsibility collectively rather than individually, and so forth. Nevertheless, many of the core issues remain. In particular, the equal shares argument suffers from the problems of climate change blinders and feasibility.

7.3.1 Climate Change Blinders

Notwithstanding its simplicity and intuitive appeal, most philosophers have rejected the equal per capita rights approach.[32] Peter Singer, whose approach was discussed in chapter 5, is the most prominent remaining proponent. The reason that most philosophers have rejected the approach is, I believe, that they recognize the climate change blinders problem. And the reason why they recognize the climate change blinders problem here is that equality claims more generally (i.e., outside the climate change context) have been criticized widely in the philosophical literature for precisely this problem.

One well-known version of the criticism is found in the writings of Amartya Sen. In his monograph on inequality, *Inequality Reexamined*, Sen notes that all theories of justice require equality.[33] The theories differ in what type of equality they demand. Egalitarian theories focus on equality of income. John Rawls would require equal liberty and equality in the distribution of primary goods. Ronald Dworkin would require equality of resources. Sen himself would require equality of basic capabilities. While utilitarians formally want to maximize the sum of individual utilities, they treat each person's utility equally in the maximization. Even Robert Nozick's libertarian theories, which on the surface seem averse to equality claims, require equality of libertarian rights.

Since every theory requires equality, the central question in pursuing justice is not whether to demand equality, it is what sort of equality to seek. That is, to understand what justice requires, we have to answer the question: equality of what? Sen's core argument is that we cannot just adopt an intuitively appealing demand for equality of something without asking why we equalize that and not something else.

Some commentators have taken this logic to mean that equality is an empty concept.[34] All of the work is being done by a theory that tells us what we should seek to equalize: resources, opportunity, income, primary goods, or whatever. Once we have agreed on this, there is no additional work for equality to do. Sen himself believes equality is still a useful concept. He argues that once we have chosen what items of value are central to justice, equality is a simple and powerful tool that helps us understand when the demands of justice are not being met. But even in this case, equality is just a heuristic. All the work is being done in the choice of what to equalize.

Regardless of our views on whether the idea of equality is a helpful heuristic or is empty, Sen's argument is precisely the climate change blinders argument. We cannot choose equality of some particular item, such as the right to use up the remaining atmospheric capacity to absorb CO_2, without a theory explaining why we want to equalize that item and not something else. Focusing only on the atmosphere is like putting on blinders to the sorts of equality that we really care about.

The blinders problem is quite general but it arises even just within the domain of climate change. The reason is that people and countries are situated differently with respect to climate change. As Stephen Gardiner notes, "people in different parts of the world have different energy needs. For example, those in northern Canada require fuel for heating that those in more temperate zones do not."[35] Countries will also differ in how much they suffer from

climate change. Some countries might be relatively imper-
vious because they are wealthy, have cool or temperate
climates, and are otherwise well situated. Others might
be especially vulnerable. And some countries, facing equal
harms, might have higher costs of reducing emissions than
others. For example, some countries may have very little
hydroelectric or geothermal power, or few places to locate
windmills or solar cells. Others might be blessed with cheap
clean energy. Equal division of emissions rights does not
produce equal outcomes, even if we just focus on climate
change. Needs, harms, and mitigation opportunities vary
widely. Even focusing just within climate policy, equal allocation of emissions rights would not produce the type of
equality most theories of justice would demand.

But the problem is even broader. There is no reason to
focus equality claims on climate change instead of some
other space of equality. As Sen notes, we need an underlying theory of justice to determine what should be equal-
ized. Backers of equal division of the atmosphere do not
offer such a theory, but almost any theory they choose
would equalize something else.

In particular, the allocation of emission rights determines the type of energy people can use and its price. It is
about money. Once we are focusing on equality of money,
however, there is no reason to want one particular resource
to be allocated equally as opposed to eliminating, or at least
reducing, inequality of wealth or income more generally.

Eliminating or reducing inequality generally, however, is the goal of distributive justice. And all of the flaws
with the distributive justice argument apply. In particular,
we should want to achieve our distributive goals in the
most effective way possible. Equal division of emissions

rights might be part of such a program, but it might not. Philosophy can tell us that distributive goals are important but not how to achieve them.

Gardiner made this argument aptly.[36] He asked how many would support the equal shares claim if the distributive consequences were to help the rich and hurt the poor? If you would not, then what you care about (correctly in my view) is the distribution of resources, not the formal division of one particular resource. But this puts us back into the problems with the distributive justice argument.

7.3.2 Feasibility

The only way left to support the equal shares claim is to argue that it is a feasible way to reduce overall inequality. Because climate negotiations will one way or another divide up the atmosphere, the allocation of rights is salient. Equal rights is a natural focal point. It is easy to understand and intuitive. Moreover, by combining equal rights with a cap-and-trade system, we can improve the distribution of resources while ensuring the emissions reductions occur where they are cheapest. Equal division of rights to emit CO_2, we might say, is a great opportunity to reduce inequality.

The problem, however, is that a treaty that allocated emissions rights this way is not feasible. Once we look at the size of the transfers and the problem with how the transfers would be made under such a system, it is evident that nations would never agree to this system.

To get a sense of the feasibility of an equal per capita emissions system, we can calculate the net transfers that would arise if emissions were at current levels.[37] To do this,

I took the emissions, population, and GDP data for 2010[38] and calculated how much each country exceeds the global per capita average emissions. Multiplying this by the population of each country gives us the total excess emissions for each country. Multiplying that by the price of permits gives the net transfer each country would have to make to purchase or sell permits so that its emissions stay the same. We do not know what price permits would trade at. The number would depend on how ambitious the reduction target is. I used the same $35 per ton that I used for the corrective justice calculation. The number could easily be lower or higher. The transfers would scale accordingly.

The transfers are massive. The United States would be the largest payer on pure dollar terms. It would have to pay the rest of the world $163 billion per year to purchase the necessary permits. This is about eight times its current foreign aid budget. The total federal budget for the US each year is about $2 trillion so the payments would be about eight percent of total government spending, and would be roughly a third the size of Medicare (which is currently about $550 billion per year). Perhaps most importantly, the net transfers would be about 2% of GDP and would be near and likely exceed the benefits that the United States would receive from stopping moderate climate change (very substantial climate change might produce larger harms to the United States).

China currently exceeds the global per capita emissions so it too would have to make payments. Its net outgoing payments would be about $32 billion per year, which is about 1% of its GDP. Russia would have to transfer $47 billion per year, which is an astounding 5.2% of GDP. Brazil would have to transfer $28 billion per year, which is 2.5%

of its GDP. The twenty-seven countries in the EU have relatively modest emissions on a per capita basis given their income: they would have to transfer $48 billion per year, but this is only 0.3% of their collective GDP.

The largest net recipient in absolute dollar amounts would be India, which would receive more than $200 billion per year, or 17% of GDP. Other large recipients include Bangladesh ($32 billion per year, 39% of GDP) and Pakistan $30 billion per year, 22% of GDP). A number of countries in Africa (Eritrea, Niger, Ethiopia, and Malawi) would receive annual transfers roughly the size of their current GDP.

Looking at these numbers, it seems clear that the United States, China, Russia, and Brazil—four out of the top five emitting countries—would never agree to this system. Compare it, for example, to a treaty in which countries agree to emissions reductions targets but can achieve them however they want. If a country achieves its target through a tax on emissions or a domestic cap-and-trade system, it would keep the revenue. It is hard to see why the largest emitters would not insist on this alternative route.

The system overall would likely be progressive because most rich countries would be net payers. It would not, however, be tailored to national income. As a result, some very poor countries would have to make substantial (or effectively impossible) net payments. Belize would have to make payments equal to 62% of GDP, Paraguay and Bolivia both would have to pay 33% of GDP each year. Uzbekistan, with per capita income of $752, would have to make payments of 5.7% of its GDP. Papua New Guinea, with per capita income of $955 would have to make payments of 4.3% of GDP. Middle-income countries such as Hungary, Croatia, Chile, and Turkey would also have to make very substantial payments.

An immediate reaction is that of course we would not ask Belize to pay 62% of its GDP as part of a global emissions rights system. It would be ridiculous to ask Papua New Guinea to further impoverish its people. But that is precisely the point. As Stephen Gardiner noted, would we actually support equal emissions rights if the distributive consequences were bad? Probably not, which is why we would not ask Paraguay and Papua New Guinea to further impoverish themselves. But if we are really pursuing distributive concerns, then we need to focus on how best to achieve those goals rather than on formal equality.

On top of these problems, the transfers would very likely go to governments not individuals. Government officials may be likely to misuse the transfers or, in many cases, appropriate them. Donor nations normally impose strict conditions on aid for this reason. If the transfers are part of a cap-and-trade system based on a theory of equal per capita rights, there would be no such limitation. If a country has a right, rooted in justice, to a resource, how could we then put restrictions on its use? But nations will not agree to transfers of this magnitude if their main effect is to enrich corrupt governments.

The equal per capita rights system continues to generate support because of its seeming fairness. But at its core, it is a claim about a feasible means of achieving distributive goals. It is not feasible, however. The most important countries for achieving emissions reductions—China, the United States, Russia, and Brazil—would all be substantial losers under this system. The transfers would go to governments who may misuse the resources. It is hard to see high-emitting countries agreeing to this. Moreover, applied strictly according to its own logic rather than as a means of achieving distributive goals, it would hurt many poor people. Despite its simple appeal, the logic behind equal per capita emissions rights is flawed.

7.4 CONCLUSIONS

Each of the major philosophical arguments about the proper shape of climate policy suffers from basic flaws. Some suffer from idiosyncratic, internal logical problems but the focus here has been on systematic problems common to all of the arguments. By thinking of the problem as part of the "ethics of climate change," people ignore the fact that climate change policy is just part of a web of policies, including policies to help the poor and policies concerning ownership of resources. They operate with climate change blinders. Removing these blinders shows that the recommendations from climate ethics do not follow from the arguments.

Moreover, the arguments tend to produce conclusions that are not within the realm of feasibility. It is sometimes good for ethical arguments to be aspirational, to ask us to stretch beyond what we think we can do. But suggestions that will simply never happen are not helpful, particularly when we face an immediate global threat.

Notes

1. Gardiner's list of the theories of justice that are most applicable to climate change is made up of procedural justice, distributive justice, and corrective justice. See Gardiner, "Climate Justice," in *The Oxford Handbook of Climate Justice and Society*, eds. John S. Dryzek, Richard B. Norgaard, and David

Schlosberg (Oxford: Oxford University Press, 2011), 310. I do not discuss procedural justice and separate equality from distributive justice. Gardiner refers to theories of equality separately on p. 315.

2. Much of the material in this chapter is taken from Posner and Weisbach, *Climate Change Justice* (Princeton, NJ: Princeton University Press, 2010).

3. For a brief overview of this vast subject, see Julian Lamont and Christi Favor, "Distributive Justice," in *The Stanford Encyclopedia of Philosophy*, ed. Edward N. Zalta, Stanford, CA: 2008, http://plato.stanford.edu/archives/fall2008/entries/justice-distributive.

4. Henry Shue, "Subsistence Emissions and Luxury Emissions," *Law & Policy* 15, no. 1 (1993): 39–59. Shue unfortunately injects concerns about animal cruelty via feedlots into the discussion, arguably confounding the appeal of the pure distributive argument with the appeal of concerns about the treatment of animals.

5. Paul G. Harris, *World Ethics and Climate Change: From International to Global Justice*, (Edinburgh: Edinburgh University Press, 2010), 131–132.

6. Stephen M. Gardiner, *A Perfect Moral Storm: The Ethical Tragedy of Climate Change* (New York: Oxford University Press, 2011), 314–315.

7. For other examples of claims from distributive justice see Henry Shue, "Global Environment and International Inequality," *International Affairs* 75, no. 3 (1999): 537–540; Peter Singer, *One World: The Ethics of Globalization* (New Haven: Yale University Press: 2002); Garvey, *The Ethics of Climate Change*, 81–83.

8. Garvey, *The Ethics of Climate Change*. London: Continuum, 2008, 871–882.

9. An additional requirement for applying any theory of justice to climate change is some form of cosmopolitanism. We have to believe that people who live in one nation owe duties to people who live in other nations, or perhaps nations owe duties to other nations or to people in other nations. While this view is contested, I will take it as correct.

10. For example, the Millennium Challenge Corporation has very specific governance requirements for nations to qualify for aid in order to ensure that the money is well spent. See Curt Tarnoff, "Millennium Challenge Corporation," CRS Report for Congress (Washington DC: Congressional Research Service, 2015), https://www.fas.org/sgp/crs/row/RL32427.pdf.

11. Shue, "Global Environment and International Inequality," 537.

12. Mathias Frisch, "Climate Change Justice," *Philosophy & Public Affairs* 40, no. 3 (2012): 230–231.

13. Darrell Moellendorf, "Climate Change and Global Justice," *Wiley Interdisciplinary Reviews: Climate Change* 3, no. 2 (2012): 131–138.

14. Gardiner, Stephen M. "Ethics and Global Climate Change." *Ethics* 114, no. 3 (2004): 579. This statement comes at the beginning of a section of his paper on responsibility for past emissions but it not clear whether Gardiner believes the statement to be supported only by a theory of responsibility or whether the consensus view is supported more generally, including by theories of distributive justice. In the next paragraph he notes a lack of consensus on backward-looking considerations, which is the core issue of responsibility.

15. Gardiner, "Climate Justice," 314.

16. Singer, *One World: The Ethics of Globalization*, 33–34.

17. See Steve Vanderheiden, *Atmospheric Justice, A Political Theory of Climate Change* (New York: Oxford University Press, 2008); Garvey, *The Ethics of Climate Change*; Harris, *World Ethics and Climate Change, From International to Global Justice*; Stephen M. Gardiner, et al., *Climate Ethics: Essential Reading* (Oxford: Oxford University Press, 2010); Eric Neumayer, "In Defence of Historical Accountability for Greenhouse Gas Emissions," *Ecological Economics* 33, no. 2 (2000): 185–192.

18. Shue, "Global Environment and International Inequality," 533–537.

19. Gardiner, "Ethics and Global Climate Change," 583.

20. Posner and Weisbach, *Climate Change Justice*, 99–118.

21. See Stephen M. Gardiner, *A Perfect Moral Storm: The Ethical Tragedy of Climate Change* (New York: Oxford University Press, 2011), 418.
22. For example, Simon Caney, "Cosmopolitan Justice, Responsibility, and Global Climate Change," *Leiden Journal of International Law* 18, no. 4 (2005): 747–775 (see p. 757: the beneficiary pays principle "is not a revision of the 'polluter pays' approach, it is an abandonment of it.")
23. Singer, *One World: The Ethics of Globalization.*
24. Ibid.
25. Gardiner, "Ethics and Global Climate Change," 582.
26. Singer, *One World: The Ethics of Globalization.*
27. I use the data from Michel den Elzen, et al., "Countries' Contributions to Climate Change," *Climatic Change* 121, no. 2 (November 1, 2013): 397–412, available for download at http://www.pbl.nl/en/publications/countries-contributions-to-climate-change. The numbers were computed by taking the difference between cumulative emissions for a country divided by the total number of lives in that country during the time period and cumulative global emissions divided by total lives during the same time period. This gives the excess per person emissions. This amount was multiplied by the total lives in the country and by the carbon price. Calculations are available from the author. The carbon price is not adjusted for time. The reason is that the carbon price should increase over time at roughly a constant rate. This means that the price in a prior year should be the current price discounted by that constant rate. But the amount would have been owed in that prior year so the amount owed today is the future value of that amount. Taking discounted values and future values at the same rate offsets, we can simply use the current price and add up the total amounts.
28. Shue, *Climate Justice*, 4; Gardiner, *A Perfect Moral Storm*, 419.
29. Paul Baer, "Equity, Greenhouse Gas Emissions, and Global Common Resources," in *Climate Change Policy: A Survey*, ed. Stephen Schneider (Washington, DC: Island Press, 2002), 393–408. Note that Baer no longer advocates for equal per capita rights. See also Paul Baer, et al., "Greenhouse Development Rights: A Proposal for a Fair Global Climate

Treaty," *Ethics, Place & Environment* 12, no. 3 (2009): 267–281; Paul Baer, "The Greenhouse Development Rights Framework for Global Burden Sharing: Reflection on Principles and Prospects," *Wiley Interdisciplinary Reviews: Climate Change* 4, no. 1 (2013): 61–71.

30. Paul Baer, et al., "Equity and Greenhouse Gas Responsibility," *Science* 289, no. 5488 (2000): 2287.

31. Michael Grubb, "The Greenhouse Effect: Negotiating Targets," *International Affairs (Royal Institute of International Affairs)* 66, no. 1 (1990): 67. Support in the philosophical literature includes Gardiner, et al., *Climate Ethics, Essential Reading*, 271–272; Sven Bode, "Equal Emissions per Capita over Time—a Proposal to Combine Responsibility and Equity of Rights for Post-2012 GHG Emission Entitlement Allocation," *European Environment* 14, no. 5 (2004): 300–316; Singer, *One World: The Ethics of Globalization*.

32. For example, as noted, Paul Baer, a prominent proponent of equal per capita rights, has since abandoned the approach. Gardiner rejects the approach for the reasons similar to those discussed in the text. See Gardiner, *A Perfect Moral Storm*, 422–423; Simon Caney, "Climate Change, Energy Rights and Equality," in *The Ethics of Climate Change*, ed. Denis Arnold (Cambridge, UK: Cambridge University Press, 2011), 10. Moellendorf, "Climate Change and Global Justice," 138–139.

33. Amartya Sen, *Inequality Reexamined* (Cambridge, MA: Harvard University Press, 1992).

34. Peter Westen, "The Empty Idea of Equality," *Harvard Law Review* 95, no. 3 (1982): 537–596.

35. Gardiner, *A Perfect Moral Storm*, 422.

36. Ibid., 423.

37. A global cap-and-trade system would generate emissions reductions, so the numbers would not exactly reflect what would happen but the reductions would very likely be modest in the short run. So this calculation gives a reasonable estimate of the transfers that would occur in the early years of such a program.

38. I used data from the World Resources Institute database, available at www.cait.wri.org. Accessed February 27, 2014.

8

Summing Up

IT IS TIME to sum up. I have tried to establish two central propositions. The first, found in chapter 6, is that it is in our self-interest to pursue aggressive emissions reductions, starting now or in the immediate future. The remaining capacity of the atmosphere to absorb CO_2 and other greenhouse gases is sufficiently small that even if we start now and move rapidly, we will have a hard time keeping temperature increases to reasonable levels. The harms, while long-term, also threaten us in the not too distant future.

One of the intuitions that climate change is an ethical problem is that it involves harms to others. When you pollute by, say, driving, taking a hot shower, or heating and cooling your home, you harm others. Our duty not to harm others can be framed as an ethical issue.

But properly understood, climate change involves the overuse of a common resource, the atmosphere. From a purely individual perspective, there is an incentive to overuse the resource, to pollute. The harms mostly fall on people who live in far-off countries or the future. The costs of not polluting fall on you. But it is in our collective self-interest to manage the resource wisely. In the end, everyone suffers if everyone is free to dump their waste into the atmosphere. We are better off if we agree to stop.

While we are better off if we agree to stop polluting, coming to an agreement to do so is devilishly difficult. Each nation has an incentive to free-ride, to let the others bear most of the burden. We are better off if we all agree not to pollute, but each nation is even more better off if everyone else stops polluting but it gets to continue. But if everyone follows this strategy, we have no agreement and we all suffer. The solution has to solve the problem of free-riding. To say that it is in our self-interest to dramatically reduce emissions of greenhouse gases does not mean that it is easy to achieve.

The second proposition, discussed in chapter 7, is that the ethical arguments that have been made about climate change add little or nothing. They suffer from basic logical problems and propose solutions that are infeasible. The core logical flaw is what I have called "climate change blinders." The arguments about how various theories of justice apply to climate change view climate change as the domain to be considered rather than as just one policy among many. For example, we face a climate change problem and a problem of vast income and wealth inequalities. Arguments about climate change view climate change policies as having to address income or wealth inequality rather than thinking of them both as problems that may be connected but may have distinct solutions. If the solutions to one of these problems are not connected to the solutions to the other, unthinkingly treating the two problems as one is a mistake.

The core ethical arguments also suggest solutions to climate change that are infeasible. They require high-emitting nations to enter into treaties that make them worse off. Agreeing to a workable treaty will be difficult even if it makes everyone better off. If an agreement will make high-emitting nations worse off, it will be impossible. And these infeasible treaties are not even necessary because they are supported only by flawed logic.

Climate change is caused by the use of fossil fuels. Stopping climate change is difficult because fossil fuels are the basis of modern society. Virtually everything we do relies on fossil fuels. Fossil fuels are so pervasive that we don't see them, except perhaps when we fill up the tank at the gas station. Take an inventory as you go about your day. The building you live in, or go to school or work in, was constructed using fossil fuels. It is heated and cooled using fossil fuels. You likely got there using fossil fuels. Your shower and meals required fossil fuels. The paper or computer screen you are reading this on requires fossil fuels. Just about everything we do relies to a great extent on energy, and the primary source of energy is fossil fuels; they are so pervasive and so reliable that they are invisible.

Conservation can buy time, but at the end of the day, the only way to solve the problem of climate change is to replace the existing fossil fuel energy system with a system that uses clean energy. We also have to provide cheap and reliable power to those who lack it, but without locking in new emissions that will push us beyond tolerable temperature increases. This is a massive task requiring engineering and science. And it needs to be done on a global basis. Finding a feasible treaty—one that sufficiently reduces emissions of greenhouse gases, that nations will agree to and comply with—is hard because of the incentive to free-ride. We need the disciplines of political science, economics, and law to help craft a workable treaty.

Ethics and philosophy more generally can bring important insights to the problem, but it does not strike me that the solutions to climate change are centrally about ethics. The solutions involve more efficient solar and wind power, better transmissions grids, batteries that hold more energy, new designs for vehicles, treaty design and enforcement mechanisms, and so forth.

Does this view take the force out of social movements to protect the planet? Will people really want to march on Washington to clamor for efficiency improvements in the electricity grid, better batteries, or more engineers? There are special interest groups who benefit from the status quo and who are able to spend vast sums to block action. Only a motivated population will be able to overcome them. Don't we stand a better chance of doing so if we frame the issue as moral, as the next great civil rights challenge? Does reducing the problem to a technocratic one doom the possibility of a solution?

Perhaps. Perhaps some philosophical arguments can be instrumental even if they are not good philosophy. I leave that to the leaders of social movements. To politicians. Regardless of how our leaders motivate us, we need to drastically cut, and eventually stop, emissions to protect ourselves, our children, and our grandchildren. Let us hope we make the right choices.

PART III

RESPONSES

STEPHEN M. GARDINER
AND DAVID A. WEISBACH

9

"The Feasible Is Political"
Stephen M. Gardiner

DAVID WEISBACH'S CALL for a "policy" approach to climate change that dismisses ethics is admirably serious, passionate, and engaging. It provides a deeply tempting account of the realities of global climate politics, especially for an affluent Western audience. Nevertheless, I believe we should largely resist this vision, since it fails to take seriously the ethical challenge of the perfect moral storm. Let me briefly highlight a few initial impressions.

9.1 DANGEROUSLY UTOPIAN?

Weisbach largely assumes away the intergenerational and theoretical storms, and especially the tyranny of the contemporary and lack of adequate intergenerational institutions. On the surface, this is done through a hardheaded assertion of the power of self-interest: climate change is about "saving our own necks." However, this rhetoric obscures an intriguingly complex view: Weisbach supposes that conventional governments *reliably pursue* the interests of their citizens; adopts a *three-generation* conception of self-interest ("ourselves, our children and grandchildren"); declares that "*even using a narrow notion of self-interest*, it

is in our self-interest to wisely govern use of the atmosphere"[1]; asserts that conventional cost-benefit analysis
with high discount rates *fully takes into account* even distant future generations[2]; and assumes that these factors
converge on the same climate policy. Unfortunately, these
claims strike me as very bold, initially implausible, difficult
to justify, and open to diverse and competing interpreta-
tions. (Notably, in this volume at least, they are largely
asserted, rather than defended; moreover, some of the
cited sources outside this volume make substantial con-
cessions to ethics.[3]) Particularly noticeable is the optimism
the intriguing view embodies about contemporary global
politics, and the theoretical robustness of market CBA.
Weisbach (rightly) worries about dangerous utopianism in
ethics; yet economic realists must address the same charge,
especially given the temptations of the perfect storm.

9.2 OUTLIER

Whereas Weisbach and I both accept the basic scientific
consensus on climate change, favor "aggressive" emissions
reductions, see political inertia from Rio to Copenhagen as
a lamentable policy failure, and endorse a precautionary
approach, this convergence put him sharply at odds with
many of his natural allies. Most neoclassical economists
reject aggressive reductions, and many realists see them
as politically infeasible. Though an outlier, Weisbach typically glosses over such internal disputes. Nevertheless, in
my view, understanding them is essential for assessing the
appeal of economic realist ideology, not least because the
roots of the internal disputes are typically ethical.

9.3 QUESTIONABLE AMBITION?

Though Weisbach's actual climate policy is couched in tough, apparently uncompromising language that will appeal to many environmentalists, I remain unsure how robust the position and arguments are. On the surface, Weisbach promises *"aggressive* reductions in emissions" (49) *"far more ambitious* than those currently on the table" (9), including zero emissions in the *"not-too-distant* future," *"even if the temperature turns out to be relatively insensitive* to greenhouse gases and *even if the harms turn out to be on the lower end* of the possibilities." Still, given the shunning of ethics, we should expect important gaps in the arguments.

Consider, for example, zero emissions "in the not-too-distant future." Weisbach concedes that he cannot give a precise date, and then goes on to argue that it is somewhere between 2020 and 2150. Yet this range is enormous, and zero in five years and zero in 135 are dramatically different policy goals. Weisbach himself plumbs for the rather conservative target of 2100,[4] suggesting "starting now and *going slowly* is the best bet."[5] Still, this is done without much explanation, and it is unclear how it results in emissions reductions *"far more ambitious* than those currently on the table" (e.g., the Copenhagen Accord targeted 50% on 1990 levels by 2050, with at least 80% in the developed countries). It is also not obvious why unusually stringent targets would be warranted or feasible. Consider, for instance, Weisbach's claim that radically fast decarbonization would be justified even if climate sensitivity turns out to be very low. Suppose "very low" means something resulting in a temperature rise of 1°C by 2100 (i.e., the bottom end of IPCC projections). 1°C is significantly below Copenhagen's

1.5°–2°C benchmarks, and many scientists would regard it as "manageable" through adaptation. Consequently, such an unusually stringent target is likely to be seriously controversial, especially among economic realists (why not "invest" elsewhere?). Moreover, although I have more sympathy for Weisbach's precautionary approach (usually regarded as anathema by CBA's partisans),[6] in my view it requires ethical underpinnings.

9.4 KYOTO'S FOLLY

Of course, Weisbach is adamant that ethics will lead us astray. His prime example comes from a 2004 article in which I noted a surprising consensus among climate ethicists that developing nations should not face emissions limits in the "foreseeable future." Unfortunately, this example strikes me as a red herring.

First, the remark is taken out of context.[7] The 2004 piece surveyed the (very small) literature of the 1990s and early 2000s. Back then, discussion focused on the fight to create and implement the Kyoto Protocol. Consequently, the policy-relevant "foreseeable future" was *the first commitment period* under the UNFCCC, which was negotiated through the 1990s, came into effect in 2004, and expired in 2012. Negotiations are now far beyond this context. From 2005 onwards, the second commitment period was negotiated, culminating in 2009's Copenhagen Accord, which covers emissions to 2020. Recent negotiations concern a third commitment period, to be agreed in Paris in 2015 and expected to run until 2030 and perhaps beyond.

Second, in negotiations for the first commitment period, the emissions exemption had many advocates, and Weisbach seems sympathetic to their reasons. Proponents not only tended to argue that it was infeasible to impose immediate restrictions on the emerging economies, but that it would be politically necessary for the developed countries, and especially the United States, to demonstrate their seriousness about cuts by going first. Such views explicitly assumed that developing countries would accept cuts in later periods (often post-2020). Weisbach acknowledges and appears to approve of this strategy.[8]

Third, ironically, my own position was much more critical. Anticipating Kyoto's failure, in 2001 and 2004 I argued that "all countries should be explicitly included in the regime," that "combating climate change requires full cooperation of at least all countries of significant size, including the United States, China, and India," that "costs must be borne by almost everyone," and climate policy should be linked to other global cooperative ventures, such as trade.[9] Consequently, I rejected the Kyoto approach, calling it a "dangerous illusion" predictable within the perfect moral storm.

Moreover, whereas Weisbach takes Kyoto to be flawed because of a preoccupation with justice, I agreed with those critics interpreting it as dominated by concerns for efficiency and a "two-track" approach, elements Weisbach continues to argue are central to future success. More broadly, his account sits awkwardly with the irony expressed by many negotiators, that ultimately "the [Kyoto] system is made in America, and the Americans aren't part of it."[10]

9.5 EVOLVING TRAGEDY

Weisbach may concede that my intergenerational storm and tyranny of the contemporary analysis can explain the Kyoto debacle.[11] Yet he insists that the present situation is different because the current generation (and its children and grandchildren) are threatened by medium-term impacts. By contrast, I argued in my book that a generation concerned with such impacts may still engage in climate policy that is unethical with respect to the further future.[12] Faced with looming threats to itself (and perhaps its nearest and dearest), it may overemphasize adaptation at the expense of mitigation, substantially increase emissions to boost production of goods that aid short-term protection, or even pursue "parochial" geoengineering techniques that aim to hold off the worst for a century or so while imposing even more severe risks on later generations. Crucially, *medium-term impacts do not guarantee even minimally decent climate policy*, and may open up avenues for more pronounced intergenerational buck passing.

9.6 EXCLUDING JUSTICE

In confronting the global storm, Weisbach proposes *excluding whole categories of concern* from climate policy (e.g., distributive and corrective justice), even despite their importance to the large emerging economies whose cooperation is essential. This strikes me as unhelpful (e.g., I doubt the political wisdom of saying to India " 'feasibility' means accepting our world view—where you are the problem and your concerns about us are irrelevant"). Notably, it under-estimates the extent to which the feasible is political.

Excluding justice is often made to look more plau-sible by arguments that posit an extreme version of a particular justice concern (e.g., past emissions), make it the sole determinant of climate policy, and then suggest problematic results (such as "some past emitters are cur-rently poor"). However, most approaches to climate eth-ics are *pluralistic* in ways that diffuse such worries (e.g., several factors matter, extreme poverty trumps historical responsibility).

In addition, the extreme arguments typically aim to frighten us with numbers, framed as demands for large transfers of cold cash. Yet these numbers are often far from robust, and the demands misleading. For instance, consid-ering past emissions, developing countries should probably be compensated most for the extent to which overconsump-tion of fossil fuels by developed countries has effectively deprived them of the same opportunity for cheap develop-ment. The most relevant costs are therefore opportunity costs. Given this, poor countries might be compensated by richer countries' facilitating development in other ways (e.g., technology transfers, promoting alternative energy, remov-ing protectionist trade practices). This is very different from handing over lump sums, and likely to be less expensive (e.g., reducing protectionism may be mutually beneficial).

Similarly, even if full compensation turned out to be unrealistic, other cases suggest that substantial recon-ciliation can still be achieved through partial redress (e.g., the Ngai Tahu settlement in New Zealand). Good faith efforts are better than simply dismissing victims' claims as "infeasible."

9.7 EFFICIENCY BLINDERS

Weisbach rejects any claim of justice that goes beyond what is "cost-effective" in reducing emissions. Yet this seems one-sided: rejecting justice can also make reductions more expensive (e.g., if, under International Paretianism, Bangladesh must pay off the big emitters). Worse, the rhetoric encourages "efficiency blinders." The ultimate point of climate policy is not elimination of emissions for its own sake, but protecting people and the rest of nature. Requiring poor farmers to compromise their fragile subsistence because it "costs less" than the rich forgoing a few luxuries risks forgetting this.

Notably, recognizing climate justice need not involve a commitment to righting all the world's wrongs through climate policy. While some may hope to open the door to a dramatically new world order (a vanguard model), others aim merely for modest improvements in areas that intersect with climate (a mild rectification model), or at not worsening wider injustice (a neutrality model). In my view, discussion of the relative merits of these models is an important topic in the ethics of the transition. Unfortunately, requiring the most vulnerable to pay in the name of "efficiency" seems in tension with all of them.

Of course, Weisbach is strangely hostile to the ethics of the transition, dismissing anything short of a "complete background understanding of international justice" as unphilosophical hand-waving. Yet this strikes me as unreasonable (especially within a theoretical storm), and also uncharitable to those of a less radically cosmopolitan bent (e.g., proponents of institutional conceptions of international justice, who want to improve the world, proceeding from where we are).

9.8 BEARING WITNESS

I am also unmoved by complaints that pointing out gross injustices is "idle." I have argued that ethics can and should provide guidance for action (even given the theoretical storm). Moreover, although one purpose of ethics is to guide action, in my view it also plays a role in *bearing witness* to serious wrongs even when there is little chance of change. Ideal theory is central to this task; but an ethics of transition can also play a part. Though we may not yet know precisely what an ethical climate would look like, concerns about injustice (for example) provide guidelines that help hold us accountable. This is so even if mostly what they do is remind us that our current behavior falls far short of any morally defensible goal. Recall that, in a perfect moral storm the key worry is not that ethics may require us to be angels, but that we may (rightly) be remembered as something far worse.

Notes

1. 149, this volume.
2. 168, note 22, this volume.
3. 167, note 18, this volume.
4. 179, this volume: "we have to reduce . . . to near zero, by sometime around the end of this century."
5. 186, this volume, my emphasis.

6. Stephen M. Gardiner, "A Core Precautionary Principle." *Journal of Political Philosophy* 14, no. 1 (2006): 33–60; Stephen M. Gardiner, "Ethics and Global Climate Change." *Ethics* 114, no. 3 (2004): 555–600

7. On substance, Stephen M. Gardiner, *A Perfect Moral Storm: The Ethical Tragedy of Climate Change*. New York: Oxford University Press, 2011; chapter 4.

8. 168, note 30, this volume.

9. Stephen M. Gardiner, "The Global Warming Tragedy and the Dangerous Illusion of the Kyoto Protocol," *Ethics and International Affairs* 18, no. 1 (2004): 23–29, 28, 39; Stephen M. Gardiner, "The Real Tragedy of the Commons," *Philosophy and Public Affairs* 30, no. 4 (2001): 387–416, 411–412.

10. David Doniger, quoted in Gardiner, "The Global Warming Tragedy," 591.

11. 171, this volume: "[climate] might have been primarily a long-term problem when negotiations began in the early 1990s."

12. Stephen M. Gardiner, *A Perfect Moral Storm*, chapters 6 and 10.

We Agree: The Failure of Climate Ethics

I WROTE MY portions of this book, as well as several related works, because of concerns that the theories of ethics, or justice more generally, that had been put forward with respect to climate policy were flawed. They suffered from internal logical problems—their conclusions failed to follow from their premises. Moreover, following these theories would lead us outside of feasible solutions to a pressing global problem, or if feasible, would fail to stop climate change. I suggested that pursuing enlightened self-interest would lead to strong global emissions reductions, with the main problem being solving the free-rider problem.

The task for those defending the use of ethics to shape climate change policy is to put forward a theory of ethics that does not suffer from internal logical problems, that can actually be implemented, and that when implemented will solve the problem. This is not a high bar. It simply asks that if we are to follow a theory, that it make sense and work.

Gardiner has chosen not to take up the task. He does not put forward or defend a theory of ethics that meets these minimal requirements. Rather, he agrees with me that existing theories fail. His core claim is we face a "theoretical storm" by which he means "we are extremely ill-equipped to deal with many problems characteristic of the long-term future." In his view:

> ... even our best theories struggle to address basic issues such as intergenerational equity, contingent persons, nonhuman animals, and nature. Climate change involves all these and more. Given this, humanity appears to be charging into an area where we are theoretically inept, in the (nonpejorative) sense of being unsuited for (e.g., poorly adapted to), or lacking the basic skills and competence to complete, the task.
>
> ...
>
> [T]he theoretical storm also afflicts major theories in moral and political philosophy, such as utilitarianism, Rawlsian liberalism, human rights theory, libertarianism, virtue ethics, and so on. I conceive of such theories as major research programs that evolve over time (rather than, say, as sets of propositions). Whatever their other merits, in their current forms these research programs appear to lack the resources needed to deal with problems like climate change. Moreover, it seems likely that in evolving to meet them, they will be substantially, and perhaps radically, transformed.
>
> Importantly, I am not claiming that *in principle* such theories have nothing to say. On the contrary, at a superficial level, it is relatively easy for the standard research programs to assert that their favored values are relevant to climate change and license condemnation of political inertia. Severe and catastrophic forms of climate change pose a big enough threat that it is plausible to claim that most important values (e.g., happiness, human rights, freedom, property rights, etc.) are threatened. Surely, then, proponents of such values will think that something should be done. Nevertheless, in my view the more important questions are how precisely to understand the threat, and what should be done to address it. On such topics, the standard

research programs seem curiously oblivious, complacent, and opaque, even evasive. In particular, so far they have offered little guidance on the central question of the kinds of institutions that are needed to confront the problem, and the specific norms that should govern those institutions. Though this situation is starting to change as theoretical attention shifts (after twenty-five years of waiting), there is still a very long way to go.

These paragraphs could have come out of my chapters.

Rather than trying to meet these minimal requirements for an ethical theory, Gardiner suggests a lower bar for understanding the role of ethics: that in principle ethics is relevant even if right now it is not up to the task of providing guidance. It is hard to disagree with this because we cannot exclude the possibility that someone might come up with a sound, feasible theory. But until we see such a theory, we cannot know what it might say and in what direction it might lead. Instead, all we can do is address the ethics-based approaches that have been put forward so far. And here Gardiner and I agree: they fail.

This of course leaves many smaller disagreements. Gardiner defends elements of corrective justice that I do not believe are valid. He is less willing to live with feasibility constraints than I am. He thinks self-interest will lead to moral corruption. I do not. And so on. I leave these disagreements to our respective chapters where we each have had our say. What is most important is that we agree on the central issue: existing theories of ethics fail. As Gardiner says, they "offer[] little guidance on the central question of the kinds of institutions that are needed to confront the problem, and the specific norms that should govern those institutions." I agree.

INDEX